ADVANCED LEVEL

Data and Data Handling for AS Level

BIOLOGY

Bill Indge

Orders: please contact Bookpoint Ltd, 130 Milton Park, Abingdon, Oxon OX14 4SB. Telephone: (44) 01235 827720. Fax: (44) 01235 400454. Lines are open from 9.00 – 6.00, Monday to Saturday, with a 24 hour message answering service. You can also order through our website www.hodderheadline.co.uk.

British Library Cataloguing in Publication Data
A catalogue record for this title is available from the British Library

ISBN 0 340 856475

First Published 2003
Impression number 10 9 8 7 6 5 4 3 2
Year 2009 2008 2007 2006 2005 2004 2003

Typeset by Tech-Set Ltd, Gateshead, Tyne & Wear.
Printed in Great Britain for Hodder & Stoughton Educational, a division of Hodder Headline, 338 Euston Road, London NW1 3BH by Martins the Printers, Berwick upon Tweed.

Contents

LIVERPOOL
COMMUNITY
COLLEGE

Introduction

A hundred years ago we knew a lot less about biology than we do today. Electron microscopes had not been invented and we had very little knowledge of the structure of cells or of the organelles present inside them (see Figure 1). In the area of genetics we were only just beginning to gain an understanding of inheritance. The discovery of the structure of DNA was still 50 years in the future, and topics such as gene therapy and genetic engineering, which now form part of your AS course, were unheard of.

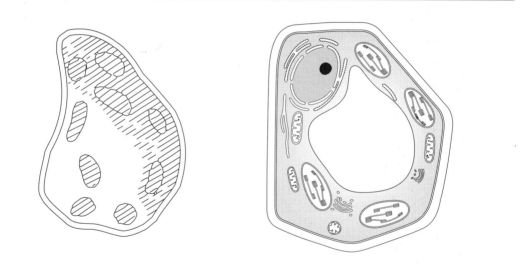

Figure 1 *The drawing on the left was taken from a textbook of botany published in 1864. It shows a chlorophyll-containing cell from a leaf. The drawing on the right shows a similar cell. It was taken from a modern textbook. We now know far more about the structure of such cells, and experiments performed by biologists and analysis of the resulting data have led to a much greater understanding of the processes that take place within them*

About this book

Many of the discoveries which advanced our knowledge in the twentieth century resulted from scientific investigations which involved well designed experiments and careful analysis of the results. These skills need to be mastered by all scientists. To be able to use a proper scientific approach to a particular problem is just as important to a biologist as is a knowledge of cell structure or an understanding of genes and DNA. The first three chapters in this book will help you to develop these experimental and analytical skills. Chapter 1 looks at the way in which we design experiments and plan investigations to collect evidence. Chapter 2 concentrates on how we can manipulate the data we collect and present them in different forms – tables, bar charts, histograms and line graphs. The theme of Chapter 3 is the interpretation of these tables and graphs. These three chapters have all been written in a similar way and have a number of features in common. These include:

- **Text questions.** The text questions are designed to help you to understand what you have just been reading and are meant to be answered as you go along. Most of them are very straightforward and can be attempted from the information in the paragraph or two which came immediately before. No answers have been included. If you get stuck, read the last paragraph again. If you still have problems, make a note of the question and ask for help.

- **Summary boxes.** Conventions are standard ways of doing things. Biologists have conventional ways of doing such things as designing experiments, constructing tables and drawing graphs. It is very important that we all follow these conventions so they have been summarized in boxes spread through these first three chapters. You should find these summaries useful in answering the questions in the remaining chapters.

- **Exercises.** Each chapter finishes with one or more exercises. These exercises contain questions based on the contents of the chapter and should help you to master the relevant skills.

There are a number of different AS Biology and Human Biology specifications. They differ in detail but they are similar in many ways. Have a look at the specification you are following. Somewhere at the front you will find a section headed 'Assessment Objectives'. This section is extremely important because it sets out the skills you will need to master and on which you will be tested. It does not matter which specification you are following, this section is exactly the same. If you read it carefully, and look at the assessment grid which follows, you will see that your unit tests will have around a third of the available marks allocated to Assessment

Objective 2. This requires you to demonstrate the ability to show that you can, among other things, 'translate data from one form to another' and 'interpret data in tables and graphs'. You can add to this some of the additional assessment objectives which form part of your coursework. Here you will see references to 'devising experiments and planning investigations', and to 'interpreting, explaining and evaluating the results' of these experiments and investigations. The exercises which make up the remaining chapters in this book are intended to help you to gain the considerable range of skills required to meet these assessment objectives.

The emphasis in all these exercises is on developing the skills associated with handling data, so it is a good idea to make sure that you have a sound knowledge of the topic concerned before you start work on a particular exercise. To help you with this knowledge, the subject matter that you will need to draw on in order to answer the questions has been written in a box underneath the title. In the earlier exercises you will also come across the word **Hint** at various places. These hints are intended to point you in the right direction. Some of them provide a little more help; others refer you to one of the boxes in the first three chapters. As you work through the exercises, you will hopefully gain in confidence. This is why there are fewer hints in the later chapters.

Before you start

The questions in this book are designed to help you to acquire a range of skills. They are not intended to catch you out but, before you start, you must take a little time to analyse the question and see what it wants you to do. There is little point in spending a lot of time on these exercises if you do not follow the instructions given in the individual questions. Analysing questions requires you to do three things:

1 **Look at the command word.** This a word such as 'describe' or 'calculate' or 'explain' which tells you what you are required to do. Box 1 contains a list of the important command words that have been used in writing the questions in this book. Use it from the start to make sure that you get into the habit of giving the required answer. The one expression that you will not find in this box is 'Write all you know about … ' We will not use it here, and nor will your examiners!

2 **Make sure you know what you are required to write about.** The rest of the question will tell you this.

3 **Note the mark allocation.** Mark allocations can be very helpful. In general, in data handling questions, one mark means you are required to do one thing or recall one piece of information that you have learnt during your AS course.

So, get into good habits now. Ask yourself before you answer any question:

- What have I got to do?
- What have I got to write about?
- How many marks are available?

It does not take very long, but it will help you to turn your knowledge and understanding into examination marks when the time comes.

BOX 1 Instructions used in data-handling questions

- **Calculate** means calculate! Although no explanation ought to be required for this term, an important point should be made. Make sure that you show your method of working as clearly as possible. In some cases it is possible to get marks for the right approach, even if the actual answer is wrong. You will only get marks, however, if you explain your method clearly enough.

- **Describe** involves giving a written description of information presented in a table or graph. All that you need to do is to summarize the main trends or patterns and relate these to the figures provided. Answers such as 'The rate of reaction increases steadily to a peak value of 4.2 $cm^3\,h^{-1}$ at 50°C. It then falls to zero at 65°C' should gain full credit. An explanation is not required.

- **Explain** means give a reason why. A description is not required and will not gain credit. Perhaps the best check is to ask yourself, 'Have I explained why ... ?' In AS examinations more marks are lost through failing to give proper explanations than for almost any other reason.

- **Give evidence from/Using examples from** involves making use of the material provided to illustrate a particular point. Since this is a requirement of the question, full marks will not be given for a general answer which fails to refer to the relevant material.

- **Sketch** is used where a curve has to be added to a graph. When this term is used, it is simply the shape of the curve that is required. There is no need to invent figures and attempt to plot these accurately.

- **Suggest** is used where it is expected that you will not be able to answer from memory. There may be more than one valid answer and, in general, any sensible response based on sound biological reasoning will be acceptable.

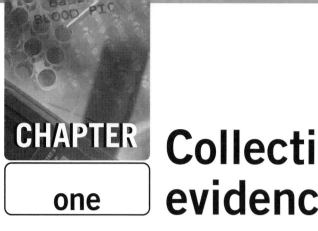

Collecting the evidence

As you work through your AS course and study topics such as cell structure and biological molecules, you might think that biology is a very theoretical subject – a mass of information that has to be learnt. Biology is much more than this. It is a science and, as such, it involves investigating problems in a particular way. Scientists make observations. These observations lead to hypotheses, and they test these hypotheses by designing and carrying out experiments. It is experiments which provide us with the data which we manipulate and analyse.

Biology is a huge subject. At one extreme it involves the study of the individual molecules that make up a living organism; at the other, we might look at how all the different organisms living in a particular ecosystem interact with each other and with their environment. In spite of this, however, almost all experiments rely on the same general approach. In this chapter we will look at the basic principles which underlie the design of biological experiments. These same principles apply whether the experiment is part of a programme of field-work or whether it is carried out in the laboratory.

At the start

In all investigations something changes or is changed by the person carrying out the experiment. This is the **independent variable**. As a result of these changes, you have to measure something else – the **dependent variable**. Your first step in designing any experiment is to identify these two variables. This will give you a clear idea of what you are trying to do. We will consider an example. Figure 1.1 shows some fruits from a sycamore tree. Each fruit has a seed at one end and a broad flat wing. When the fruits fall from the tree in the autumn, they spin round and round. This slows the speed at which they fall to the ground and enables them to be blown considerable distances from the parent tree.

Suppose we investigate the effect of the surface area of the wing on the time a fruit takes to fall to the ground. The surface area of the wing is the independent variable

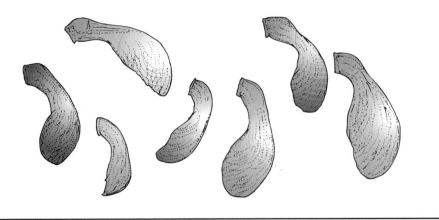

Figure 1.1 *Sycamore fruits. The wing on the fruit causes it to spin and fall slowly to the ground*

because this is the variable which is under the control of the person carrying out the experiment. The dependent variable is the time taken for the fruit to fall to the ground. The experiment will therefore be based on selecting fruits with different wing areas and timing their fall. We will change the independent variable and measure the dependent variable.

Q1   An investigation was carried out to find whether tomato juice contained a substance which, in certain concentrations, prevented the germination of lettuce seeds. In this investigation what was
    (a) the independent variable?
    (b) the dependent variable?

Changing the independent variable

An experimental result which occurs only once could be due to chance. It could also be a peculiarity of the way in which a particular biologist carried out the investigation concerned. Valid experiments must be capable of being repeated by someone else. In order to make this possible, clear details must be provided. This means that you should provide enough detail so that another person in your group can follow your technique exactly as you intended, without the need for any help other than your set of instructions. Some aspects of your method may be very obvious. It is quite enough, for example, to say that you will maintain the temperature at a constant 30 °C by using a water bath. You do not need to draw a diagram of a water bath or explain how it is set up. Other features may not be so clear. You will not get away with statements like, 'Put some glucose in a test tube'. What exactly do you mean? What do you mean by 'some'? Are you going to use solid glucose or is it a solution? If it is a solution, what concentration are you going to use? It is essential to include details. It would be much better to say 'Put 5 cm^3 of glucose solution in a test tube. The concentration should be 0.2 mol dm^{-3}.' Then someone else can follow your method exactly.

In most experimental situations, other variables than the one you are planning to change might also affect the outcome. These are called **confounding variables** and the experiment should be designed so that their effect can be eliminated. Let us consider again our investigation into the effect of the surface area of the wing on the time a sycamore fruit takes to fall to the ground. There are several possible confounding variables. As the surface area of the wing increases, so will the mass of the fruit. Wind speed and height from which the fruit is released could also account for the differences in the time taken to fall to the ground. Clearly something must be done about these confounding variables. The simplest solution is to design our experiment so that they remain constant. If the mass of the fruit, wind speed and the distance the fruit falls are kept constant there is a reasonable degree of certainty that any difference in the time taken to fall to the ground can be explained by differences in surface area of the wing.

Q2   Does the concentration of tomato juice affect the germination of lettuce seeds? List three confounding variables which might affect the outcome of this investigation.

The control group

Once you have decided how you are going to change the independent variable and identified the various confounding variables whose values must keep constant you need to ask another question, 'Is there anything, other than the independent variable, in the way that I have set up this experiment that could have produced these results?'. If there is, you need a **control**.

A **control experiment** is one that you set up to eliminate certain possibilities. Coffee contains caffeine. Suppose you wish to investigate whether drinking coffee has an effect on heart rate. You may decide to compare the heart rates in a group of people drinking coffee with those of a group who have gone without. The control group would be the group that had gone without coffee. But, think for a minute. Would this really be a good control? An increase in heart rate in the coffee drinkers might be due to the effect of caffeine but it could simply be the result of drinking the large amount of water that was used to make the coffee! What you should do is to treat the people in the control group by giving them an equal volume of hot water to drink ... even better, an equal volume of decaffeinated coffee. By setting up this control you have eliminated the possibility that it was something other than the caffeine that had the effect on heart rate.

Q3   Cabbage rootflies are insect pests. The female rootflies lay their eggs in the soil at the base of young cabbage plants. The larvae hatch and burrow through the soil. They feed on the roots of the cabbage plants. An investigation was carried out to see whether a particular type of insecticide reduced the damage caused by cabbage rootflies. Discs of felt soaked in a solution of the insecticide were placed round the cabbage plants. Explain which of the following would be the better control experiment for this investigation

A   Leaving the cabbage plants in the control group untreated.
B   Placing discs of felt soaked in water round the cabbage roots.

Measuring the dependent variable

When you have decided how you will change the independent variable in your investigation, you should consider how you will determine the effect of this on the dependent variable. You need to collect quantitative data and this will involve taking measurements. These measurements must be made with the appropriate degree of precision and must be reliable.

Let us consider an example. Anthocyanin is a red substance found in the cytoplasm of beetroot cells. If you place a piece of beetroot in water, the anthocyanin comes out of the beetroot into the surrounding water. Suppose you wanted to investigate the effect of temperature on the amount of red dye coming out into the water. It would be no good just measuring the amount of dye coming out, you would have to specify a time period as well or you could not compare the effect at different temperatures. You have a problem making the measurements because you are dealing with a very small quantity of anthocyanin – so small that it would be impossible to separate it from the water and weigh it. The best approach would be to measure the redness of the solution after a fixed time. The redder it is the more anthocyanin has come out. How do you measure redness? There are several possibilities. You could use a colorimeter. This is an electronic instrument which passes a beam of light through the solution we are interested in and allows us to measure the amount that has been absorbed. The more light that is absorbed, the more pigment is present. There are other, simpler methods however. You could make a series of standard solutions, each one more dilute that the one before. Your unknown solution could then be compared with these standards. Or you could make a single standard solution by crushing up a piece of beetroot in water and timing for how long you needed to incubate your piece of beetroot until the surrounding solution was the same colour as the standard.

Q4 Which of the three methods of measuring the dependent variable would you expect to give the most accurate results? Explain your answer.

You also need to bear in mind that if you only carry out the experiment once, you have no idea whether or not the results are reliable. Carrying it out twice does not help you very much either. If the two results differ, which one will you take as being reliable? You need to carry out the experiment several times, each time doing it in exactly the same way. Only in this way can you separate reliable data from anomalies.

Look at Box 2. It contains a summary of some of the important things you should always consider in designing an experiment. You should use it as a checklist.

BOX 2 | **Designing an experiment**

At the start

■ Write out a title to your investigation. It should take the form
'The effect of [independent variable] on [dependent variable].'
This will give you a clear idea of what you are setting out to test.
■ Use this title to identify what you are going to change (the independent variable) and what you will measure as a result (the dependent variable).

The independent variable

■ Make sure that you have described your method of changing the independent variable in enough detail so that another person in your class could carry out the investigation exactly as you intended.
■ Consider the range of values of the independent variable which you need to consider in order to provide a comprehensive set of results.
■ List the other variables that might influence the results. Describe how these confounding variables should be kept constant.
■ Ask yourself whether anything else in the way the experiment has been set up, other than the independent variable, could have produced the results you might expect. If there is, you need a control.

BOX 2 continued

The dependent variable

■ Explain how you will measure the dependent variable and obtain quantitative results.

■ Specify the number of times you will repeat each set of readings. The more repeats you have, the more reliable your results are likely to be. Unfortunately, the more repeats you carry out, the longer the investigation will take. You will obviously need to compromise.

Exercise 1.1 What goes in must come out

Here is a simple problem. A diuretic is a substance which increases the amount of urine produced. How could you investigate the diuretic effect of drinking tea?

First steps

1 In this investigation what is

    (a)   the independent variable?

    (b)   the dependent variable?

(2 marks)

Changing the independent variable

2 Table 1.1 shows some factors which ought to be kept constant in this investigation.

Table 1.1 Factors to be kept constant in the investigation

Factor	Why the factor should be kept constant
Food eaten before	Salty food reduces the amount of urine produced by altering the salt concentration of the blood
Physical activity	
Time of day	
Body mass	

    (a)   For each factor in the table, complete the second column to suggest why it is necessary to keep it constant.

(3 marks)

    (b)   Suggest **one** more factor that should be kept constant and give an explanation for this.

(2 marks)

Exercise 1.1 *continued*

The control group

You now need to ask: 'Is there anything, other than the independent variable, in the way that I have set up this experiment that could have produced these results?' If there is, you need a control. Suppose, for example, you decide to compare the volume of urine produced by a group of people drinking tea with that of a group who had gone without. An increased volume of urine produced by the tea drinkers might be due to the diuretic effect of the tea.

3   What else might explain the increased volume of urine produced by those drinking tea?

(1 mark)

4   Which of the following would be the best control? To give the control group

      A   nothing to drink

      B   an equal volume of another drink such as coffee

      C   an equal volume of hot water

      D   an equal volume of hot water with the same amount of milk as in the tea.

(1 mark)

5   Now describe how you will change the independent variable in this investigation.

- Describe what you intend to do in sufficient detail for someone else to be able to carry out the investigation exactly as you intended.
- List the other variables that might influence the results and decide how to keep them constant.
- Describe any controls you would use.

(6 marks)

Exercise 1.1 *continued*

Measuring the dependent variable

Once you have changed the independent variable, you must determine the effect of the changes on the dependent variable. To do this you need to take measurements. Not only do you need quantitative data but you must ensure that these data are collected with the appropriate degree of precision. It is no good here, for example, simply measuring the volume of urine produced. You need to specify a time period as well.

6　Why is it necessary to specify a time period over which urine is collected?

(2 marks)

7　Describe how you will measure the dependent variable in this investigation. Remember that measurements relating to the dependent variable must be quantitative, reliable and reflect the right degree of precision.

(3 marks)

Exercise 1.2 Malaria – a vaccine?

Many of the diseases found in tropical countries are caused by parasites. One of the most common of these diseases is malaria with a worldwide total of approximately 3000 million cases a year. This disease is caused by tiny one-celled organisms which live and multiply inside human red blood cells. The damage they do is responsible for the symptoms of the disease – high fever and severe headaches. Malaria can also lead to anaemia and damage to various organs, brought about by blocked blood vessels. As a result there are around 2 million deaths from this disease every year, many of them young children. Malaria is spread by the bites of mosquitoes which breed in stagnant water.

Vaccines have been developed for many diseases and they are very effective. Could we develop an effective vaccine for malaria? If so, many lives could be saved. In the early 1990s, a team of scientists led by Dr Manual Pattaroyo developed a vaccine in their laboratory which they thought might be successful. Before they could be certain, they needed to test it and this involved designing a suitable experiment.

The team selected a group of riverside villages in Colombia for their investigation. Most of the people who lived here farmed small patches of land surrounded by forest. Each family lived in a wooden house with an open veranda on which they slept at night.

1   Explain why a group of riverside villages was a suitable area for the investigation.

(2 marks)

The team examined the villages. Anyone who was ill or pregnant was not allowed to take part in the investigation. Those selected were then divided into two groups, **A** and **B**. People in one of these groups were given the vaccine. The other group was treated as a control. The investigation was planned as a double-blind trial. This means that at this stage in the investigation neither the team of scientists nor the people taking part in the investigation knew who was in the experimental group and who was in the control group. They only found out when all the results were collected in.

Exercise 1.2 *continued*

2 (a) Explain why it was necessary to have a control group in this investigation.

(2 marks)

(b) Suggest how the control group would have been treated.

(2 marks)

3 Neither the team of scientists nor the people taking part in the investigation knew who was in the experimental group and who was in the control group. Suggest why it was important that

(a) the team of scientists did not know.

(2 marks)

(b) the people taking part in the investigation did not know.

(2 marks)

The people in the two groups were examined at monthly intervals to see whether or not they had malaria. The results are shown in Figure 1.2.

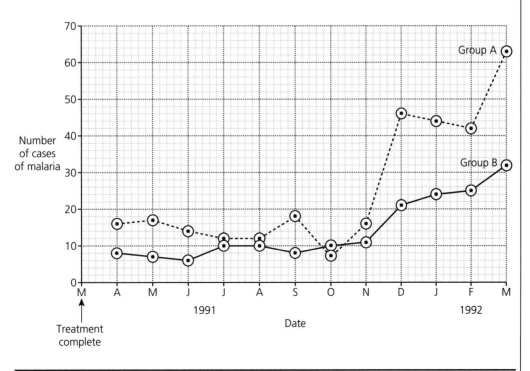

Figure 1.2 *Number of cases of malaria in control and experimental groups*

Exercise 1.2 *continued*

4   What was the dependent variable in this investigation? Give the reason for your answer.

(2 marks)

5   The research team were able to conclude that this investigation showed that the vaccine had been successful. Use this information to identify groups **A** and **B**. Give the reason for your answer.

(3 marks)

Playing around with data

Figure 2.1 shows some data. It is a very small part of a computer printout showing details about the patients attending a hospital clinic. This particular clinic deals with problems associated with bones and joints and these data relate to the first time that the patient concerned was seen by a doctor at the clinic.

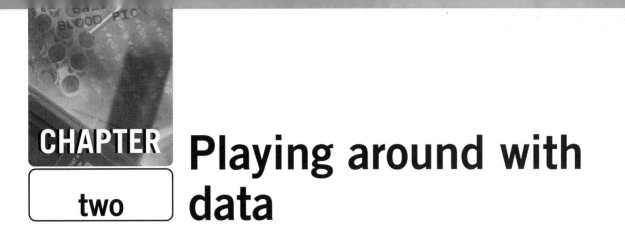

Description of condition	Age on admittance	Sex	Number of cases
Disorder of muscle or ligament	42	M	1
Fracture of ankle	42	M	1
Unspecified back problems	42	M	4
Unspecified joint problems	42	M	2
Traumatic amputation of finger	42	M	1
Gout	43	M	1
Deformity of toe	43	M	1
Disorder of muscle or ligament	43	M	1
Fracture of ankle	43	M	1
Osteoarthritis and similar conditions	43	M	3
Unspecified back problems	43	M	1

Figure 2.1 *Data from a hospital clinic. The complete printout was much longer than this – 26 pages containing information relating to 2692 patients!*

The data we actually collect as the result of a survey like this, or from an investigation in the laboratory, are called 'raw data'. One of the problems with raw data is that there is often so much information that it is very difficult to identify trends or patterns. Before information can be really useful, we need to play around with it or process it. One of the first steps we can take is to summarize the data in a suitable table, and this often means grouping items of information together. Table 2.1 is a summary of all the information provided by the printout. Look at it carefully.

You should be able to draw some obvious conclusions because you can see patterns beginning to emerge. For example

- Fractures of arm bones are most common in the 0–19 age group, particularly among boys.
- More women than men reported to the clinic with deformed toes.
- Arthritis is more frequent in older people.

Q1   Give two other conclusions that you can draw from Table 2.1.

Table 2.1 Table showing numbers of patients, separated according to age and sex, reporting with different conditions to a hospital clinic

Condition	Male				Female			
	Age/years				Age/years			
	0–19	20–39	40–59	60–79	0–19	20–39	40–59	60–79
Deformity present at birth	4	0	1	0	14	0	0	0
Fracture of arm bone	49	12	5	2	13	3	1	7
Fracture of upper leg bone	3	3	1	10	2	2	1	32
Fracture of lower leg bone	7	9	4	0	9	2	5	4
Fracture of ankle	7	8	6	2	1	7	5	6
Deformed toes	0	2	6	5	4	17	39	25
Arthritis	0	11	46	83	0	4	49	167
Unspecified joint problems	13	52	52	33	6	45	72	71
Unspecified back problems	3	56	74	48	5	65	141	64
Other conditions	54	117	184	97	62	107	148	147
Total	140	270	379	280	116	252	461	523

Constructing tables

Biologists use tables a lot and they use them in different ways. We have just seen how they can be used for summarizing data. They can also be used for making comparisons. We can use a table, for example, to compare the ways in which molecules of different substances differ from each other. In this chapter, however, we will confine ourselves to looking at tables such as Table 2.1 which contain numerical information.

If you look at a number of different biology textbooks, you will see that tables are presented in a variety of different ways. Sometimes each number is in a separate box. On other occasions, lines are only used to separate the rows of figures from each other. Units may be found in the body of the table, or they may be confined to the headings of rows and columns. This variation comes about because the job of drawing tables is given to a designer who is trying to produce pages with an attractive layout. To us as biologists, however, it is obviously much more helpful if tables are constructed in the same way. In other words, we need to follow a set of conventions. At A-level these conventions have been set out for us by the Institute of Biology. The big advantage of adopting these conventions is that all examination boards have agreed that this is the way in which they will present tables. All tables in this book also follow these conventions. They are summarized in Box 3.

BOX 3 | **Constructing tables**

■ A full grid should be drawn with lines separating both the rows and the columns.

■ The first column contains the independent variable. This is the variable for which you have chosen the values or the condition you want to investigate. In Table 2.1 we have picked out the various medical conditions, so they go into the first column.

■ The only things that should be in the body of a table of this sort are numbers. The units must be taken out and written in the headings of the rows or columns of the table. We always separate the unit from the feature with a slash or solidus as in Table 2.1. If we use compound units then we make use of negative indices. Metres per second is written as m s^{-1} and grams per cubic centimetre as g cm^{-3}. We will be using this way of writing units throughout this book. You should try to do the same.

Q2   Catalase is an enzyme. It breaks down hydrogen peroxide to produce water and oxygen. In an investigation, the volume of oxygen produced in one minute was measured when different concentrations of hydrogen peroxide were added to a fixed amount of catalase. The results were put in a table. What should have gone in the first column of the table?

Drawing graphs

Look back again at Table 2.1. It contains a lot of information, so much that it is quite difficult to see any obvious pattern. Biologists plot graphs to show patterns that are not obvious in tables. Three types of graph are used and it is important that we use the correct graph in any particular situation.

What sort of graph?

When we plot a graph, we plot values of the dependent variable against values of the independent variable. You met these terms in Chapter 1. The **independent variable** is the variable you have chosen to alter. In Table 2.1 we chose to collect data about different sorts of injury and disease. This is the independent variable. The **dependent variable**, as its name suggests, depends on the variable we have altered. In this case it is the number of patients. If you are not sure about which is the dependent variable and which is the independent variable try completing this sentence:

The [*insert the dependent variable*] depends on the [*insert the independent variable*].

If your completed sentence makes sense, you have identified these variables correctly; if it does not make sense, try again!

> Q3   A record was kept of the number of elephants visiting a water hole
> and the rainfall.
> Which of these two is the dependent variable?

With any graph, the values of the independent variable are plotted on the horizontal or **x-axis**. The values of the dependent variable go on the vertical or **y-axis**. If both sets of values are numerical, the key thing to do in deciding what sort of graph you should plot is to look at the dependent variable. If the values of the dependent variable are discrete, you should draw a bar chart or a histogram. **Discrete variables** are those which only involve whole numbers. You cannot have fractions. Examples include the number of ladybirds found on different species of plants or the number of people in different age groups.

> Q4   Is the dependent variable in Table 2.1 discrete?

We draw a bar chart or a histogram when the values on the dependent variable are discrete.

Bar charts

Bar charts are drawn when the independent variable does not have a numerical value. Suppose you were investigating the effect of different methods of cooking on the amount of vitamin C in potatoes. You should present your results as a bar chart because the independent variable is the method of cooking and this does not have a numerical value. Box 3 shows you how to draw a bar chart.

BOX 4 Drawing a bar chart

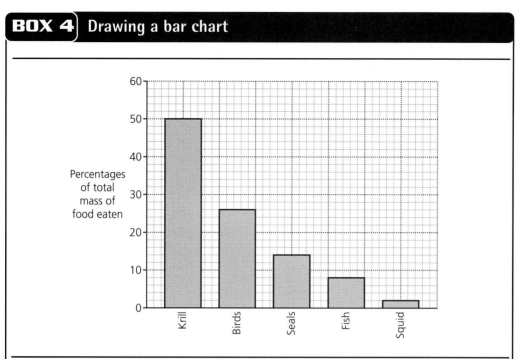

Figure 2.2 A bar chart showing the food of leopard seals in the Antarctic

- Bar charts are made up of blocks or bars of equal width. These blocks do not touch each other.
- The independent variable is plotted on the x-axis and the dependent variable on the y-axis.
- The axes should be fully labelled. Someone should be able to look at the bar chart and know exactly what it shows without any further explanation.
- There should be a title explaining what the bar chart shows.

Histograms

When the independent variable is numerical and continuous from one extreme value to the other, you should draw a histogram. The independent variable is plotted, as usual, on the x-axis, but you will need to think about the number of categories into which you group your data. This will determine the number of bars you have on your histogram. Look at Table 2.2 on the next page.

Into how many categories should we divide these marks? Suppose we made each category just two marks, as in the table. We would end up with a histogram with 27 bars. It would be very difficult to see any obvious pattern. Let us go to the other extreme. Suppose we grouped the marks into two categories, 18–45 and 46–71. It would tell us very little as any important pattern would be hidden. We need something in between. A useful rule to calculate the number of categories is to take five times the log of the total number of observations. In this case there are 75

Table 2.2 The number of students gaining particular marks in a biology unit test

Marks	Number of students	Marks	Number of students	Marks	Number of students
18 – 19	1	36 – 37	10	54 – 55	2
20 – 21	0	38 – 39	6	56 – 57	2
22 – 23	0	40 – 41	2	58 – 59	2
24 – 25	2	42 – 43	6	60 – 61	2
26 – 27	2	44 – 45	5	62 – 63	2
28 – 29	5	46 – 47	4	64 – 65	0
30 – 31	4	48 – 49	2	66 – 67	2
32 – 33	2	50 – 51	6	68 – 69	0
34 – 35	2	52 – 53	4	70 – 71	0

observations and the log of this is 1.88 (you can find the log of a particular number very easily with a scientific calculator). Five × 1.88 is 9.4, so we should try to divide the range of marks into nine equal categories. This works out at six marks in each category. Box 5 shows you how to draw a histogram.

BOX 5 Drawing a histogram

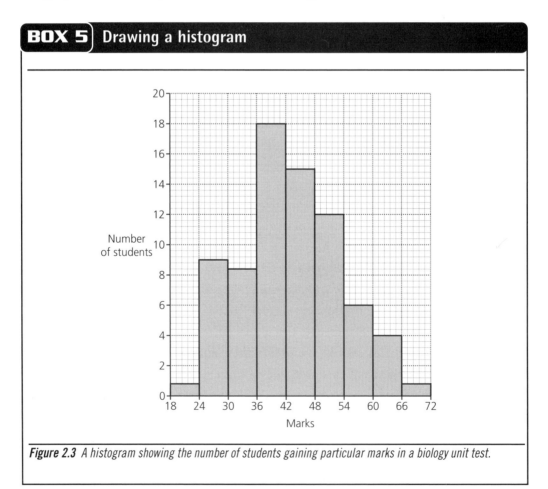

Figure 2.3 *A histogram showing the number of students gaining particular marks in a biology unit test.*

BOX 5 continued

- Start by deciding the number of categories into which you will divide the values of the independent variable. This will determine the number of blocks that you will have on your histogram. A useful rule to calculate the number of categories is to take five times the log of the total number of observations.
- The *x*-axis should represent the independent variable and the blocks drawn on a histogram should be touching. Choose an appropriate scale and mark out the *x*-axis. Mark it to show the edges of the blocks. Label these marks with the appropriate values (see Figure 2.3).
- Label the *x*-axis. Remember to separate units from the quantity that has been measured with a slash or solidus. If you use compound units you should make use of negative indices.
- The dependent variable should be plotted on the *y*-axis. In a histogram it represents the number or frequency. This axis should also be fully labelled.
- There should be a title explaining what the histogram shows.

Line graphs

You will see a lot of line graphs in your AS Biology course. Line graphs are drawn when the dependent variable is continuous. In other words, we are not dealing with whole numbers only but we have a situation where intermediate values are possible. We can have fractions. When you measure the rate of reaction of an enzyme, for example, you may record the time taken for the reaction to be completed or the volume of gas given off. Both of these are continuous variables. You can have fractions of both the minutes in which you measure the time and the cubic centimetres in which you measure the volume of gas given off.

Q5   Explain why you would use a line graph to plot the results of an investigation in which you measured the change in length of pieces of potato soaked in sucrose solutions of different concentrations.

At the start of this section you may remember that we explained that the main purpose of a graph is to show relationships which cannot be seen clearly from a table. Look at Table 2.3. It shows some of the results collected from an investigation involving the enzyme pectinase. Pectinase breaks down a substance called pectin which helps to bind the walls of plant cells together. This enzyme is used in the food industry to release more juice from fruit such as apples and oranges. The table shows the effect of adding different concentrations of pectinase on the volume of juice extracted from crushed apples.

Table 2.3 The effect of pectinase concentration on the volume of juice obtained from crushed apples

Concentration of pectinase solution/%	Volume/cm³ of juice collected after					
	0.5 hour	1 hour	2 hours	3 hours	4 hours	5 hours
0	4.1	4.6	5.7	6.2	6.5	6.6
20	4.3	5.0	6.4	8.1	8.9	9.5
40	4.8	5.8	7.6	9.8	10.4	11.3
60	4.6	6.0	8.2	9.3	11.2	12.1
80	5.0	6.4	8.7	10.0	12.1	13.0
100	6.0	7.1	10.1	12.4	13.6	14.0

A graph would obviously help us to see much more clearly the relationship between the volume of juice collected and the concentration of pectinase added. What should we plot? We need to think clearly about the aim of our experiment. We were investigating the effect of pectinase concentration on the volume of apple juice collected. The pectinase concentration is the independent variable and the volume of juice is the dependent variable. These are the two quantities that we should be plotting. Look at Figure 2.4 and the reason why we plot these two should be obvious.

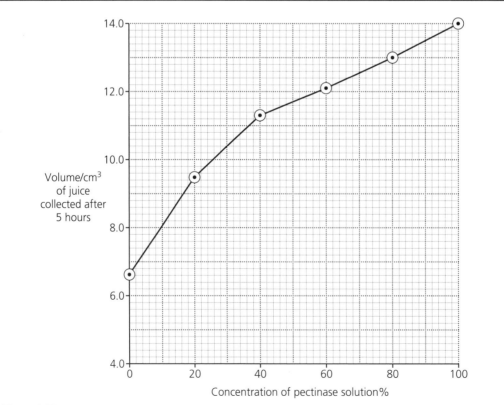

Figure 2.4A

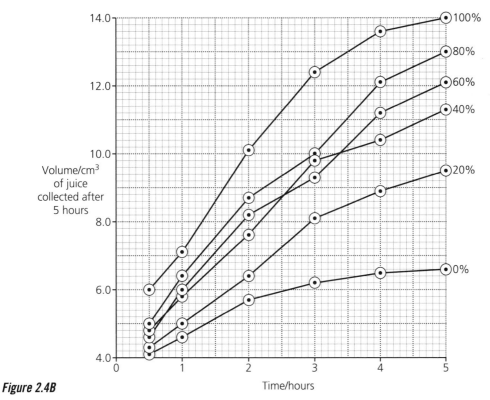

Figure 2.4B

Figure 2.4 *Line graphs should show the dependent variable plotted against the independent variable as in Figure 2.4A. Plotting all the data as a family of curves as has been done in Figure 2.4B doesn't give us a clear picture of the relationship between concentration of pectinase and the volume of juice collected.*

BOX 6 Drawing a line graph

■ The independent variable is plotted on the *x*-axis and the dependent variable on the *y*-axis.

■ The axes should be fully labelled and the units given. Someone should be able to look at the graph and know exactly what it shows without any further explanation.

■ Choose a suitable scale.
  - Make sure that all the points you need to plot will fit on the graph.
  - Avoid a scale which involves fractions of grid squares. This makes it difficult to plot points accurately.
  - Look carefully at the data and think about the range of values you need to plot. In Figure 2.5, for example, you will see that the body temperature of the camel goes from approximately 34 °C to 41 °C. If your scale went from say 0 °C to 100 °C, the curve on your graph would end up almost as a straight line. It would be very difficult to see the fluctuations in body temperature.

BOX 6 continued

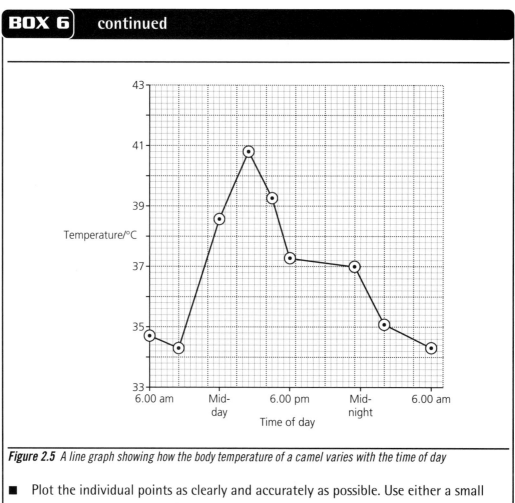

Figure 2.5 *A line graph showing how the body temperature of a camel varies with the time of day*

- Plot the individual points as clearly and accurately as possible. Use either a small dot with a circle round it (⊙) or a cross (✗).
- Join the points. Draw either a smooth curve or straight lines joining the points. A smooth curve should be used only if your data are sufficiently reliable that you feel you can confidently predict intermediate values. Otherwise, join the individual points with straight lines as shown in the graph in Figure 2.5.
- There should be a title explaining what the graph shows.

Exercise 2.1 Plague

You have probably never heard of *Yersinia pestis*. It is the bacterium which was responsible for the great outbreaks of plague which swept through Europe in the Middle Ages. The Black Death was the worst of these and killed almost a quarter of the entire population of Europe, while the Great Plague of 1665 had a terrible effect on London and other British cities (Figure 2.6).

Figure 2.6 *Burying the dead during the Great Plague in 1665. Detailed records made at the time have allowed us to follow the course of the disease in detail as it swept through London*

The data given below come from *A Journal of the Plague Year*, a book written by the novelist Daniel Defoe and published in 1722. He took the facts and figures about the number of people dying and built a novel round them. Although the book is fiction, the figures that it contains are accurate.

	Dead from all diseases	Dead from diseases other than plague
From 15 to 22 August		1331
From 22 to 29 August	7496	1394
From 29 August to 5 September	8252	1264
From 5 to 12 September	7690	1056
From 12 to 19 September	8297	1132
From 19 to 26 September	6460	927

Exercise 2.1 *continued*

The numbers of some particular articles of distempers

	Aug 8–15	Aug 15–22	Aug 22–29	Aug 29–Sept 5	Sept 5–12	Sept 12–19	Sept 19–26
Fever	353	348	383	364	332	309	268
Spotted fever	190	166	165	157	97	101	65
Surfeit	87	74	99	68	45	49	36
Teeth	113	111	133	138	128	121	112
Total	743	699	780	727	602	580	481

1 Use these figures to construct a single table showing the causes of death in London in August and September 1665. Before you start you will need to look carefully at the figures and make sure that you understand the information they show. In particular you will need to think about the following:

  • The weeks for which the data are complete.
  • What information you need to work out before you complete your table.
  • How to combine this material into a single table.

(4 marks)

2 The figures about the causes of death shown here are not as reliable as figures about causes of death collected today. Suggest **two** reasons for this.

(2 marks)

3 (a) Explain how you would calculate the percentage of deaths due to plague in a particular week from Defoe's data.

(2 marks)

  (b) Explain why knowing the percentage of deaths due to plague allows us to make a better comparison of the effects of different outbreaks of the disease.

(2 marks)

Exercise 2.2 Maggots and murder

When an animal such as a rabbit is knocked down and killed by a car, flies soon find its body. They lay their eggs on it. The eggs hatch into maggots which burrow into the body. The maggots grow rapidly and then moult to form pupae. A new generation of flies emerges from these pupae.

To a fly, a dead human is no different from any other dead mammal. Flies are just as likely to lay eggs on the body of a murder victim abandoned in a lay-by as on a dead rabbit. Forensic scientists make use of this to estimate the time of death. They must take into consideration, however, that the time taken for a fly to complete its life cycle varies with temperature. The life cycle is completed much faster in hot summer conditions than it is in cooler spring and autumn weather. Look at Table 2.4. This shows the time taken for eggs to hatch at different temperatures.

Table 2.4 The effect of temperature on the time taken for fly eggs to hatch

Temperature/°C	Time taken for eggs to hatch/h
5	230
10	90
11	70
13	50
17	30
19	25

1 Plot the data in the table as a suitable graph. Join the points with a best-fitting smooth curve.

(4 marks)

2 Use your knowledge of enzymes to suggest an explanation for the shape of the curve in your graph.

(2 marks)

Exercise 2.2 *continued*

3   A body was found half-hidden among some bushes. There were many fly eggs on it and some of these had just hatched. Use your graph to estimate how many hours had passed between the body being dumped and it being found if the temperature at the place where it was found was

(a)   15 °C

(b)   24 °C.

(2 marks)

There is a problem, however, in estimating the time of death in the way you have just done. How do we know the local temperature at the time the body was lying undiscovered? It is this temperature which determines how long the fly eggs take to hatch. What forensic scientists do is to measure the temperature at the place where the body was found and plot this against the temperature at the nearest meteorological station at which records are kept. Figure 2.7 shows some figures obtained from an actual investigation involving the body described in question 3.

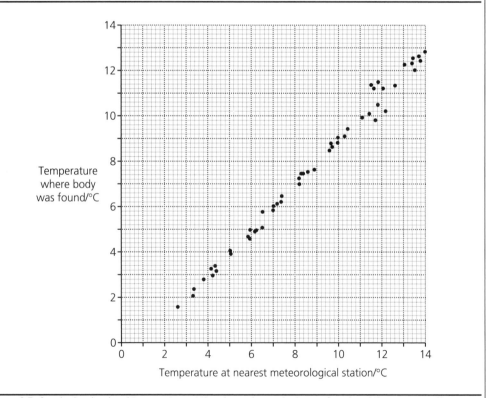

Figure 2.7 *Graph showing local temperature at the place where a body was found and the corresponding temperature at the nearest meteorological station*

Exercise 2.2 *continued*

4 Describe the relationship between the local temperature at the place where the body was found and the temperature at the nearest meteorological station.

(2 marks)

5 (a) How could a forensic scientist use the graph you have drawn and the graph in Figure 2.7 to provide an estimate of the time when the body was dumped in the bushes?

(3 marks)

(b) Explain **one** way in which using the time taken for fly eggs to hatch might give an unreliable estimate of the time at which the body was dumped.

(2 marks)

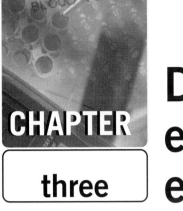

CHAPTER three

Describing, explaining and evaluating

Once we have collected data, we will need to analyse them. In this chapter we will look at the three steps involved in analysing data.

- **Describing** trends and patterns.
- **Explaining** these trends and patterns in terms of biological knowledge.
- **Evaluating** and assessing the reliability of results and conclusions.

Describing data

Red kangaroos are found in the hot, dry interior of Australia. They are very well adapted and have a number of ways in which they can stop their body temperature from increasing in these conditions. One of these is by sweating. Figure 3.1 on the next page shows the results of an investigation into sweaty kangaroos! More accurately, as the title tells us, it shows the effect of exercise on the rate of sweating.

What does this graph tell us? Let us start by looking for the basic pattern. Forget about all the small fluctuations, it is the main trends that we are interested in. You can see that during a period of exercise the rate of sweating increases. Once exercise stops, it falls again. We can link this overall pattern to particular figures. The resting level of sweat production is around $25 \, cm^3 \, m^{-2} \, h^{-1}$ and it rises to a peak value of about $200 \, cm^3 \, m^{-2} \, h^{-1}$. With this information we can produce a more detailed description of the way in which exercise affects the rate of sweating in red kangaroos:

> During a period of exercise, the rate of sweating increases from a value of $25 \, cm^3 \, m^{-2} \, h^{-1}$ to $200 \, cm^3 \, m^{-2} \, h^{-1}$. Once exercise stops, the rate of sweating falls to its resting value.

In the graph in Figure 3.1, the overall pattern is quite easy to see. In others, things are not quite so obvious. Look at Figure 3.2. This is a different sort of graph called a scatter diagram. It shows the different varieties of a common British snail, *Cepaea*

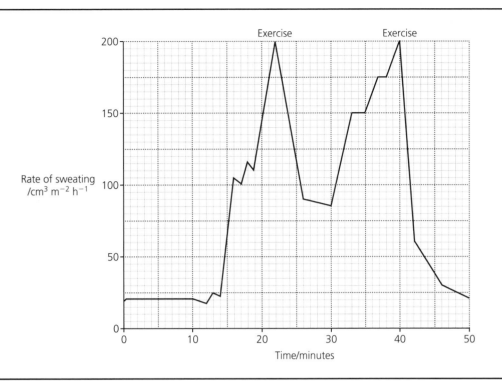

Figure 3.1 *The effect of exercise on the rate of sweating in a red kangaroo*

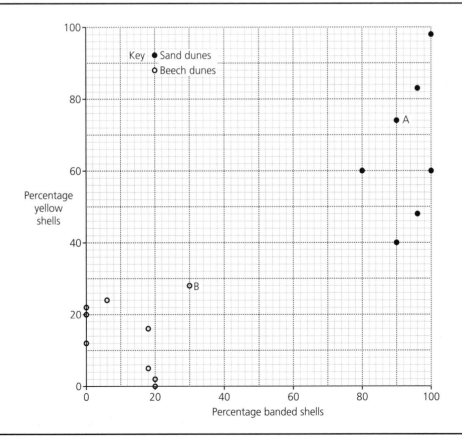

Figure 3.2 *The distribution of different varieties of the snail Cepaea nemoralis in sand dunes and beech woods*

nemoralis, found in different habitats. *Cepaea nemoralis* has a shell which can be yellow, pink or brown. Whatever the colour, the shell may have one or more black bands around it (banded shells) or it may not have any of these bands (unbanded shells).

Clearly, before we start to describe patterns, we must make sure that we understand what has been plotted. One helpful trick is to look at a specific point on the graph and try to explain exactly what it represents. Look at the point labelled **A**. It is a solid black spot so it represents a sample of snails collected in sand dunes. The *x*-axis shows us that in this sample 90 per cent of the snails had banded shells. The *y*-axis tells us that whether the shells were banded or not, 74 per cent of them were yellow in colour.

Q1   What percentage of shells in sample B were:
           (a) yellow?
           (b) pink or brown?
           (c) banded?

Q2   Describe what the graph tells us about the snails in sample B.

This is quite a good technique and it can also be used when you are trying to make sense of information in a table. In this case you need to look at the figures in a particular row or column.

Box 7 summarizes the features you should bear in mind when describing information in a graph or a table.

BOX 7 Describing data

- Start by making sure you understand the data concerned. Read the title carefully. If it is a graph, make sure that you understand the labels on the axes. If it is a table, make sure that you understand the headings of the rows and the columns. Then check your understanding by describing a particular point on the graph or column or row in the table.
- Concentrate on patterns and trends. Do not worry about minor fluctuations. With tables it is often helpful to sketch a quick graph. If a pattern exists, it will show up more clearly.
- Relate patterns and trends to values given in the table or the graph.

Mean and standard deviation

We will now look at the raw data from which graphs are plotted. Soya beans are an important crop in parts of southern Africa. An investigation was carried out into the effect of altitude on crop yield. Sites were selected at different altitudes and, at each site, ten sample plots were planted with soya beans. These grew and were harvested. Table 3.1 shows some of the results of this investigation.

Table 3.1 Yield of soya beans in plots at different altitudes in Zimbabwe

Site	Altitude/metres	Yield of soya beans in each plot/tonnes per hectare									
A	1506	3.7	4.0	4.1	4.0	3.6	4.1	3.8	4.0	4.1	3.8
B	1338	4.4	4.4	4.5	4.3	4.1	4.5	4.4	4.4	4.1	4.3
C	1292	3.7	3.4	3.3	3.2	3.5	3.2	3.2	3.1	3.3	3.2
D	1157	3.3	3.0	2.8	3.1	3.0	3.0	3.0	3.0	2.7	2.8
E	992	3.0	3.0	3.1	2.8	2.9	2.9	2.9	2.9	3.0	2.7
F	881	2.0	2.1	2.3	2.3	2.4	1.6	2.2	2.0	1.9	2.0

It is difficult to see any obvious pattern when you look at this table. The information it contains needs simplifying in some way. A good starting point is to get some idea of the middle value of each set of plots. From a mathematical point of view, there are various ways in which this can be done. The most useful to a biologist is to calculate the **mean**. You can do this for site **A** by adding up all the individual figures for yield and then dividing the total by 10, the total number of plots at site **A**, or, if you have a calculator with a statistical mode, you can use that. The mean soya bean yield for site **A** is 3.9 tonnes per hectare. Now, look again at the results for this site. You will see that the yields of the different plots vary. They range from 3.6 to 4.1 tonnes per hectare. It would be useful to have a measure of how spread out the individual values are about the mean. We call this measure the **standard deviation** (**SD**). It can be calculated from a formula but, if you have a suitable calculator, use this instead. It is much simpler and all you need to be able to do in an A-level Biology course. Table 3.2 shows the results from this same investigation but this time presented as the mean and standard deviation for each site.

We can plot the information in Table 3.2 as the graph shown in Figure 3.3. It shows the mean yield of soya beans plotted against the altitude. **Error bars** have been added for each value. Each error bar represents one standard deviation above and one standard deviation below the mean. So Figure 3.3 contains a lot of information. It not only shows us how the yield of soya beans varies with altitude but also shows us how much variation there is between the plots at each site.

Table 3.2 Yield of soya beans in plots at different altitudes in Zimbabwe. Yields are given as mean and standard deviation

Site	Altitude/metres	Yield of soya beans in each plot/tonnes per hectare	
		Mean	Standard deviation
A	1506	3.9	0.2
B	1338	4.3	0.1
C	1292	3.3	0.2
D	1157	3.0	0.2
E	992	2.9	0.1
F	881	2.1	0.2

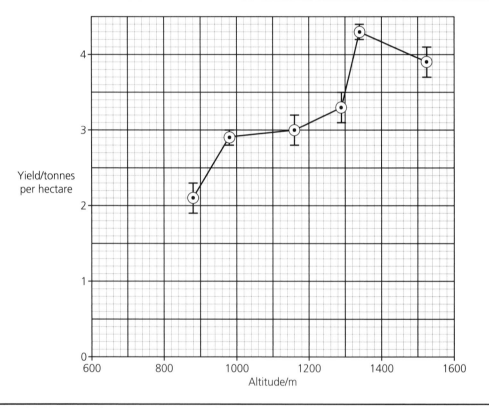

Figure 3.3 *Mean yield of soya beans in plots at different altitudes in Zimbabwe. The error bars represent one standard deviation on either side of the mean*

Q3   Use the graph in Figure 3.3 to describe how the yield of soya beans varies with altitude.

Q4   At what altitude or altitudes is the variation between the yield of soya beans in the different plots least.

Looking for relationships

All living organisms respire and most of them need oxygen to do this. There are many factors which affect the rate of oxygen consumption. One of these is temperature. Different organisms consume different amounts of oxygen at different temperatures. Biologists investigating this want to know if there is an **association** between oxygen consumption and temperature. Once they have collected the necessary data, the first thing they might do is to plot a scatter diagram with oxygen consumption on one axis and temperature on the other. Look at Figure 3.4. It shows scatter diagrams for an insect, the Colorado beetle, and a chipmunk which is a small, squirrel-like mammal.

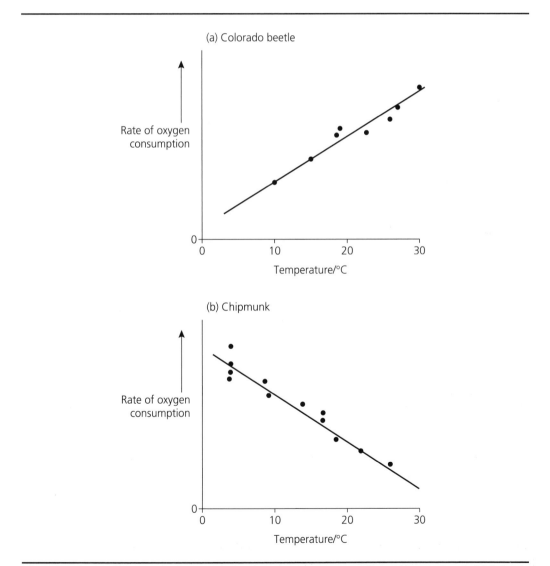

Figure 3.4 *Scatter diagrams showing associations between oxygen consumption and temperature in (a) Colorado beetle and (b) chipmunk*

Two different types of association are shown here. With the Colorado beetle, there is a **positive association**. In other words, as the temperature increases, so does the rate of respiration. A line of best fit will slope upwards. The scatter graph for the chipmunk, on the other hand, shows a **negative association**. As the temperature increases, oxygen consumption decreases, so a line of best fit slopes downwards. There is a third possibility not shown here. There may be **no association** in which case the points will be scattered randomly over the graph. It would not be possible to draw a line of best fit, or the line would run parallel to one of the axes.

Explaining patterns

Flick through the pages of this book and look briefly at the exercises it contains. You will see many tables and graphs concerned with different aspects of biology. Each set of data that you encounter during your AS or A-level Biology course is likely to be different, so you obviously cannot offer the same explanation every time! What you can do, however, is to identify a number of general points which should always be bourne in mind when suggesting an explanation. These points are listed in Box 8.

BOX 8 **Explaining patterns**

1 Take your time. Make sure you understand the data and identify obvious patterns before you start explaining.

2 The curves on many graphs are quite complex in shape. It often helps to break a complex curve down into its separate parts and look for an explanation of each part.

3 Explain means **give a reason**. It does not mean describe. If you are required to explain something, make sure that you really do give a reason.

4 You should be explaining a particular set of data so you must make sure that you have related your knowledge to the figures concerned.

We will apply these principles to a specific example. You may be familiar with the enzyme, amylase. This enzyme is found in saliva. It is also produced by fungi and by germinating seeds. It catalyses the chemical reaction in which starch is broken down to a sugar, maltose. Specific concentrations of amylase and starch were mixed and incubated at different temperatures. At 15 second intervals, samples were withdrawn from the mixture and tested for starch. The graph in Figure 3.5 shows the time taken for all the starch to disappear at different temperatures.

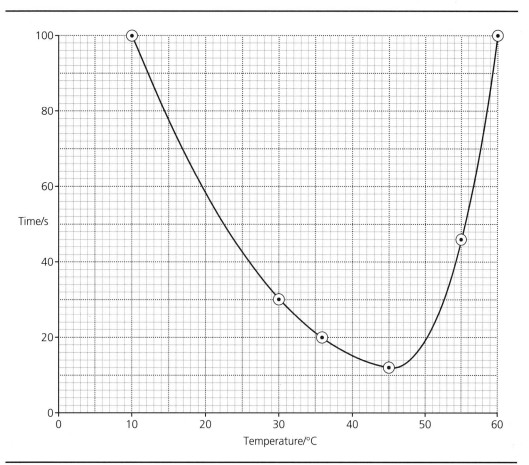

Figure 3.5 *The time taken for starch to disappear when starch and amylase were incubated at different temperatures*

If you have studied GCSE Science, you will probably be familiar with this enzyme, and should understand how temperature affects the rate of an enzyme-controlled reaction. We will look at the graph and show how the points listed in Box 8 may help to produce a comprehensive explanation of the shape of the curve. We will consider each of these points in turn.

1   The curve shows us how the rate of the reaction changed although its inverted shape may be unfamiliar. It is particularly important, therefore, that we understand the data before we attempt to explain them. At 10 °C, it takes 100 seconds for all the starch to disappear and be broken down to maltose. At 30 °C, it disappears much faster; it only takes 30 seconds. The rate of reaction is much faster at 30 °C than it is at 10 °C.

2   This curve clearly has two parts. It falls between 10 °C and 45 °C and then rises sharply. We should explain both of these aspects.

3   Now the explanation. We have to **give a reason** why the curve falls and why it rises sharply …

4   … and we need to relate this to the time taken; we shouldn't just write about the rate of the reaction.

Here is an explanation for the first part of the curve:

> Between 10 °C and 45 °C, the time taken for the starch to disappear gets less because the rate of reaction increases. As the temperature increases, the molecules of starch and amylase gain more kinetic energy. They therefore move faster and are more likely to collide and react.

Q5   **Give an explanation for the shape of the curve above 45 °C in the graph in Figure 3.5.**

Some curves will have shapes that you will meet frequently during your biology course. Figure 3.6 shows some results from a survey on ice cream sales and illustrates a particularly important point.

It is clear from this graph that there is positive association between ice cream sales and the incidence of sunburn – the greater the number of ice creams sold, the more cases of sunburn there are. We have to be very careful how we explain this. Obviously ice cream does not cause sunburn. The most likely explanation is that a third factor, temperature, is involved; the hotter the day, the greater the number of ice creams sold and the higher the number of cases of sunburn. The important thing to note is that two things may be associated but this does not necessarily mean that one causes the other.

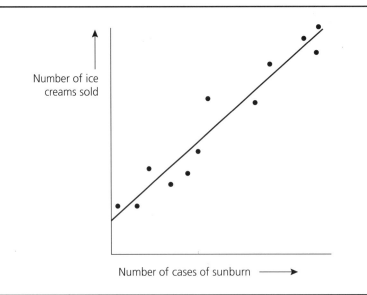

Figure 3.6 *Number of cases of sunburn plotted against ice cream sales*

Investigating the risk factors associated with disease can be quite tricky because of this. One particularly interesting example concerns the parasitic disease, schistosomiasis. This is a common disease in many parts of the tropics. People suffering from schistosomiasis are generally very lethargic and spend a lot of their time just sitting about. A study was carried out with seasonal workers on a very large sugar cane plantation. From the results of this study, a graph was plotted of the amount of sugar cane cut by each worker against the number of parasites in the worker's blood. The curve on this graph was the same shape as that in Figure 3.6. This came as a complete surprise to the research scientists because it appeared to show that having schistosomiasis made cane cutters more efficient! The real explanation was that a third factor was involved – previous experience. More experienced workers could cut more cane; they also had more parasites which they had picked up from the area round the plantation. Schistosomiasis did not make people more efficient at cutting cane after all.

We will now look at an example of another curve whose shape you will encounter frequently. This example has nothing to do with biology. Supporters are going to a football match. To get into the ground, they have to pass through turnstiles. In the graph in Figure 3.7, the rate at which supporters get into the ground has been plotted against the number of people outside trying to get in.

The curve has two main parts, labelled **A** and **B**. We will look at part **A** first. You can see that the rate of entry to the ground is directly proportional to the number of

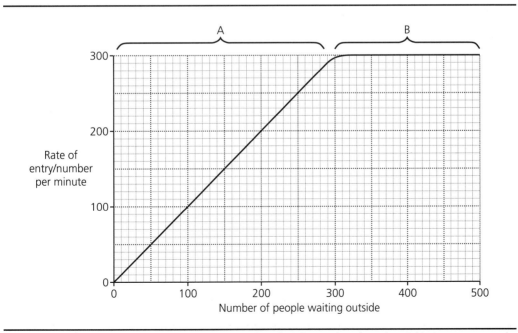

Figure 3.7 *The rate of entry to a football ground plotted against the number of people outside*

people outside. Three hours before the game starts very few people are trying to get in. They can walk straight up to a turnstile and enter. As more and more people arrive, the rate of entry to the ground increases. There comes a point, however, where there are so many people outside that all the turnstiles are working as fast as possible and queues start to build up. The rate of entry to the ground cannot get any faster. We are now on the part of the curve labelled **B**. We say that over part **A** of the curve the number of supporters trying to get in is the **limiting factor**. It limits the rate of entry into the ground. The curve levels out in part **B**. It does not matter how much faster supporters arrive at the ground, the rate of entry stays the same. Something else is acting as the limiting factor. It is probably the number of turnstiles. We will now look at a biological example of the same principle. The graph in Figure 3.8 shows how light intensity affects the rate of photosynthesis of a plant.

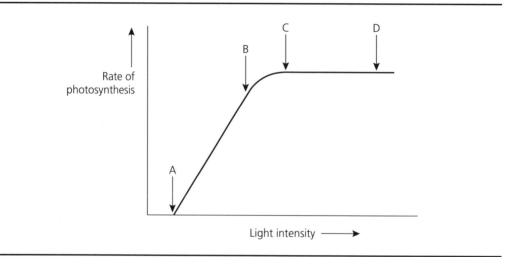

Figure 3.8 *The effect of light intensity on the rate of photosynthesis*

Q6 What factor limits the rate of photosynthesis between points A and B on the graph in Figure 3.8? Explain the evidence from the graph which supports your answer.

Q7 Suggest a factor which might limit the rate of photosynthesis between points C and D.

Watch for curves which have this shape. They are very common in biology and the explanation relies on the same principles every time.

Evaluation

We are always making comparisons and drawing conclusions – my mobile phone is better than yours ... Manchester United is the best football team ... the MMR vaccine is dangerous. Unfortunately, many of these comparisons and conclusions are subjective. They are based on personal opinions and data that we cannot really trust. As scientists, we ought to look very carefully at our work. We need to **evaluate** it and make sure the data we collect are reliable and the conclusions we draw can be trusted.

In this section we will concentrate on evaluation of investigations and their results. All investigations have three main steps. We start by designing and planning the experimental work we need to do. We then carry it out and collect the raw data. Finally we analyse the data and draw any conclusions we can.

Let us start by thinking about the first two steps – planning and carrying out the experimental work. As A-level biologists, we will have learnt a lot about experimental design and our plan should be as good as we can make it. It is important to appreciate that evaluation is not about making a list of personal failings such as 'the thermometer should have been read more accurately', or 'I should have taken more readings'. If you could have worked more accurately and reliably, why didn't you? Evaluation is concerned with identifying the limitations which cannot be avoided if particular apparatus or techniques are used.

We will consider an actual investigation to show precisely what we mean. Figure 3.9 shows how the effect of light intensity on the rate of photosynthesis was investigated.

What might produce unreliable results? The lamp might not have been the exact distance from the pond weed or the bubbles might not have been counted accurately. The point about these two things is that the lamp should have been placed the exact

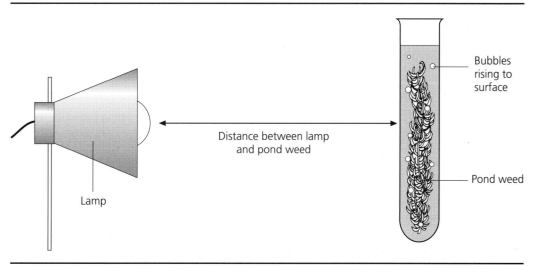

Distance between lamp
and pond weed

Lamp

Bubbles
rising to
surface

Pond weed

Figure 3.9 *The effect of light intensity on the rate of photosynthesis of pond weed*

distance away and the bubbles should have been counted accurately. Both features were within the control of the person carrying out the investigation so should not have caused a problem. These are not the sorts of things we are interested in evaluating in this investigation. There are other concerns. The lamp is likely to heat the contents of the tube and the rate of photosynthesis will be faster at higher temperatures. In addition, the bubbles of oxygen given off will not necessarily be leaving the cut end of the stem at the same rate, or be of the same size. These are the sorts of things about which we should be commenting. They are key sources of error resulting from the limitations of the apparatus and technique.

We need to do rather more than identify problems. There are four questions which we should ask ourselves.

1   **How will these sources of error affect our results?**
    A good understanding of the way in which sources of error affect our results should help us to design ways of minimizing their influence.

2   **Which of these sources of error is the most important and which the least?**
    Clearly we do not want to waste a lot of time trying to address sources of error which are not very important. We should look first at those which will have the greatest effect on the results we get.

3   **What could be done to minimize the influence of these sources of error?**
    Obviously, anything we can do here will help to make the results more reliable. In the investigation we have been discussing, we could, for example, put a small glass tank of water between the lamp and the plant. This will act as a filter and absorb some of the heat from the lamp.

4   **What other investigations could be carried out that would provide us with more information on which to base our conclusions?**
    Remember that we are particularly concerned about making our conclusions more reliable. So, if you think about the investigation we have just described, we could look at collecting the gas concerned and measuring its volume … or we could investigate the effect of temperature on the rate of photosynthesis. In this way we could see if the heating effect of the lamp did influence the results. What is not much use is to carry out a completely different experiment which would tell us nothing more about the conclusions we have drawn from our original investigation.

Exercise 3.1 Evaluating experiments

1   In winter, small birds such as blue tits and great tits spend most of the daylight feeding. This is particularly important in very cold weather as they require a lot of energy to make up for heat lost overnight.

The relationship between the mean daily temperature and the mass of peanuts eaten by small birds visiting a garden peanut feeder was investigated. At exactly the same time each morning, the mass of nuts eaten in the previous 24 hours was measured together with the mean temperature for this period. The peanut feeder and some of the results obtained are shown in Figure 3.10.

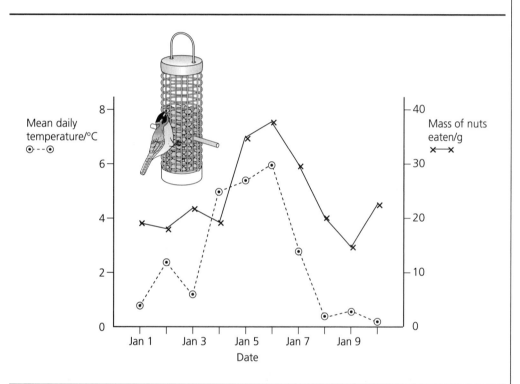

Figure 3.10 *The effect of mean daily temperature on the mass of peanuts eaten by small birds visiting a feeder*

(a)   Explain how rain may have affected the reliability of the results obtained in this investigation.

(2 marks)

(b)   The balance used in this investigation weighed to the nearest gram. Would there have been any advantage in using a more accurate balance? Explain your answer.

(1 mark)

Exercise 3.1 *continued*

2 Figure 1.1 on page 2 shows some fruits from a sycamore tree. When these fruits fall from the tree, they spin round and fall slowly to the ground. This means that the wind can blow them a long way from the parent tree. Several things influence the time taken for a particular fruit to fall to the ground. These include

- the mass of the fruit
- the surface area of the wing
- the height from which the fruit is released.

The effect of wing surface area on the time taken to fall to the ground was investigated. Fruits were taken and the surface area of the wing of each was measured by drawing round it on a sheet of graph paper. Each fruit was numbered with a felt pen. Fruits were then dropped from a height of 5 m and the time each took to fall to the ground was recorded. Each fruit was dropped five times in all and the mean value calculated.

(a)   Describe how you could make sure that the mass of the fruit would not influence the results.

(2 marks)

(b)   (i) What do you think is the greatest source of error in this investigation? Give a reason for your choice.

(2 marks)

(ii) Using the same apparatus, suggest how you could make the effect of this variable as small as possible.

(1 mark)

Exercise 3.1 *continued*

3   When a maggot moves, it wriggles. Each time it wriggles, the small black structure at its front end moves forward then back. By counting how often this happens in a given period of time, its rate of wriggling may be found. Figure 3.11 shows how the effect of temperature on the rate of wriggling of a maggot was investigated.

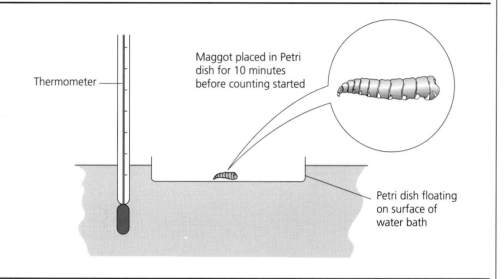

Figure 3.11 *Investigating the rate of wriggling of a maggot*

Table 3.3 shows some of the results obtained.

Table 3.3 The effect of temperature on the rate of wriggling of maggots

Temperature/ °C	Rate of wriggling/wriggles minute^{-1}
11	46
16	92
22	148
27	212
34	206

(a)   Give **two** features of this investigation which would have had an important effect on the reliability of the results.

(2 marks)

(b)   Which of these two features do you think was more important? Give a reason for your answer.

(2 marks)

Exercise 3.2 Counting rabbits

In many parts of the world rabbits are pests. They damage crops and compete with sheep and cattle for food. Because of this, a lot of research has been carried out on rabbit numbers. Rabbits usually live together in underground burrows called warrens. They feed in the open in areas around the warren. A few rabbits, however, live above ground all the time. They are called surface-dwelling rabbits.

Table 3.4 shows some counts of rabbits seen by an observer when travelling by car in Britain and New Zealand.

Table 3.4 Some observations made on rabbit numbers in Britain and in New Zealand

	Britain 1977	New Zealand	
		North Island 1965	South Island 1964
Number of live rabbits	396	5	3
Number of dead rabbits	435	20	19
Distance travelled/km	3680	3520	3840

1 Suggest a way of giving the results which would take into account the different distances travelled in the three areas.

(1 mark)

2　(a)　What are the main conclusions about numbers of rabbits that you could draw from these data?

(2 marks)

　　(b)　Suggest why you would need to be cautious in drawing conclusions from the data in this table.

(2 marks)

Exercise 3.2 *continued*

In another investigation, an observer made a regular rail journey. On each occasion, he recorded the number of rabbits seen on the north side of the railway line from a window seat facing the direction of travel. Table 3.5 shows the results for April 2002.

Table 3.5 The number of rabbits observed during a regular rail journey during April 2002

Date	Time at start	Number of rabbits seen
12 April	08.23 16.26	14 21
16 April	08.25 16.27	26 65
22 April	08.22 16.32	26 20
24 April	08.26 16.18	15 49
26 April	08.23	8
27 April	16.26	33

3   (a)   Does the time of day make any difference to the numbers of rabbits seen? Support your answer with evidence from Table 3.5.

(2 marks)

          (b)   Use the figures in Table 3.5 to suggest why it is important to collect data on as many days as possible.

(3 marks)

Exercise 3.2 *continued*

Table 3.6 shows the results of a third investigation and gives the numbers of male and female rabbits trapped at a number of sites.

Table 3.6 The age and sex of rabbits trapped inside a warren and in the open area around it

Site of trapping	Approximate age of rabbits	Number of males	Number of females
Inside warren	Under 6 weeks	140	142
	6 – 10 weeks	124	158
	11 – 16 weeks	202	265
	Mature (over 16 weeks)	2526	3357
In the open	Mature	219	132

4   Use your knowledge of X and Y chromosomes to explain why you would expect equal numbers of male and female rabbits under six weeks old.

(3 marks)

5   (a)   Describe the trend in the ratio of male to female rabbits trapped in the warrens as the rabbits get older.   *(1 mark)*

         (b)   Suggest an explanation to account for this observation.

(1 mark)

Exercise 3.2 *continued*

The home range of a rabbit is the area in which it is usually found. In an investigation in Australia, rabbits were fitted with radio-transmitters which enabled the positions of individual animals to be recorded. Figure 3.12 is a map showing the home ranges of warren-dwelling rabbits and surface-dwelling rabbits.

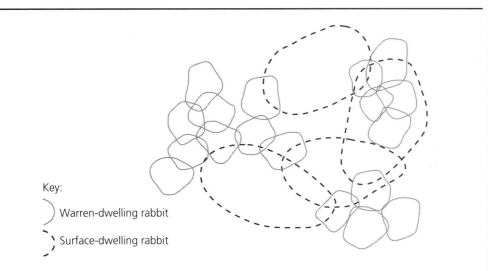

Key:

) Warren-dwelling rabbit

) Surface-dwelling rabbit

Figure 3.12 *Map showing the results of an Australian study of the home ranges of warren-dwelling and surface-dwelling rabbits*

6 Explain how this technique would enable a research worker to

(a) find out whether a particular rabbit was surface dwelling.

(1 mark)

(b) find the area of a particular rabbit's home range.

(2 marks)

7 Rabbit calcivirus disease (RCD) is a fatal disease of rabbits. In some areas of Australia, RCD reduced rabbit numbers by over 90 per cent. In other areas the same disease only reduced rabbit numbers by 50 per cent. Suggest how this difference may be explained in terms of the abundance of surface-dwelling rabbits.

(2 marks)

CHAPTER four

Cells and cell biology

Exercise 4.1 Cells and cell organelles

In order to answer the questions in this exercise, you must be able to

- recognise the main organelles present in an animal cell and have some knowledge of their functions.

Although most animal cells have a single nucleus, the number of other organelles they contain differs from one type of cell to another. This difference in number is often linked to the function of the cell. One way in which we can compare the numbers of particular organelles in different cells is to estimate their volume density. Look at Figure 4.1. This drawing has been made from an electron

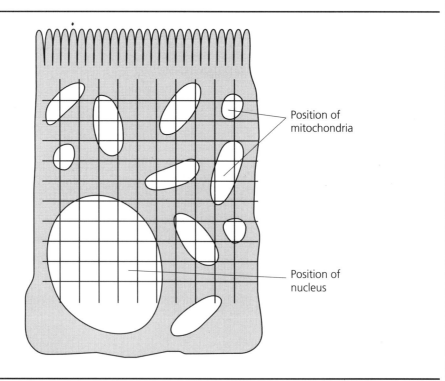

Position of mitochondria

Position of nucleus

Figure 4.1 *A cell from the lining of the small intestine showing the positions of the nucleus and the mitochondria*

micrograph of a cell from the lining of the small intestine. The only organelles shown are the nucleus and the mitochondria. They have been shown in outline only.

A grid has been drawn on this diagram. The gridlines intersect at a number of points. The volume density of the mitochondria in this cell may be calculated from the formula:

$$\text{volume density} = \frac{\text{number of points which touch mitochondria}}{\text{total number of points}} \times 100$$

1 Use the equation to calculate the volume density of mitochondria in the cell shown in Figure 4.1.

(1 mark)

2 Mitochondria produce ATP (adenosine triphosphate). This substance is an important source of energy for cells. Use your knowledge of the functions of different cells to suggest why each of the following has a high volume density of mitochondria:

(a)   a human sperm cell

(b)   a muscle cell

(c)   the cell shown in Figure 4.1

(d)   a cell from the pancreas.

(4 marks)

Hint   To answer question 2, you may need to find out a little more about the main function of these cells in your textbook.

Exercise 4.1 *continued*

3 Suggest the value of the volume density of mitochondria in a red blood cell. Give a reason for your answer.

(2 marks)

There are various techniques we can use to find out more about the functions of different organelles in a cell. One method is to use radioactive substances. The radioactivity acts as a label and it can be detected wherever it is. Figure 4.2 is a drawing of an electron micrograph of part of another animal cell. The main function of this cell is to secrete enzymes.

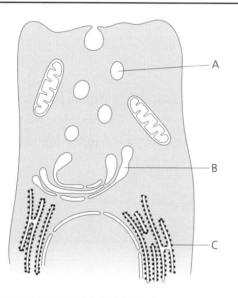

Figure 4.2 *A drawing showing part of a cell whose main function is to secrete enzymes. Letters have been used to label various organelles*

Cells like those shown in Figure 4.2 were grown in a culture. Radioactive amino acids were added to the solution in which they were growing. At various times, samples of the cells were removed from the solution. The amount of radioactivity in different organelles in these cells was measured. The results are shown in Table 4.1.

Exercise 4.1 *continued*

Table 4.1 The amount of radioactivity in different cell organelles after radioactive amino acids were added to the surrounding solution

Time after radioactive amino acids were added to the solution/minutes	Amount of radioactivity/arbitrary units		
	Golgi apparatus	Ribosomes	Vesicles
1	21	120	6
20	42	68	6
40	86	39	8
60	76	28	15
90	50	27	28
120	38	26	56

4 What happens to the amino acids on the ribosomes?

(1 mark)

5   (a)   Draw a line graph to show these results. When you have plotted the points on the graph, join them with straight lines.

(4 marks)

(b)   Explain why:

(i) the results were plotted as a line graph.

(ii) the points were joined with straight lines.

(2 marks)

Hint   If you need help with drawing your line graph look at Box 6 on page 21.

6 Use the information from your graph and the letters on Figure 4.2 to write down a sequence showing the path of radioactivity into, through, and out of the cell.

(1 mark)

Exercise 4.2 What makes a living organism

In order to answer the questions in this exercise, you must

- know about the functions of proteins, lipids and carbohydrates in living organisms.

All living organisms are made up of organic substances such as carbohydrates, proteins and lipids. These substances are called organic because they consist of large carbon-containing molecules. Organisms also contain substances such as water which have small molecules. The actual amount of water in an organism is very variable but, on average, it is about 80 per cent of the total mass. Finally there are inorganic ions such as those of calcium and sodium which are needed for the organism to function.

Table 4.2 shows the amounts of some substances in decomposing biological materials.

Table 4.2 The proportion of different substances in some decomposing biological materials

Decomposing material	Amount as percentage of total dry mass				
	Lipid	Starch	Cellulose	Protein	Ash
Dead leaves	8	26	17	9	6
Bullock carcass	50	0	0	39	11
Mammal faeces	3	6	31	8	9

1 (a) Explain why the figures for dead leaves do not add up to 100 per cent.

(1 mark)

(b) Dry mass is the mass remaining after all the water has been removed. Explain why the figures in the table are given as a percentage of the total *dry* mass.

(1 mark)

Exercise 4.2 *continued*

2 The dry material left after the water has been removed can be heated strongly and burnt. What remains after burning is called ash.

    (a)   Explain why there are no carbon-containing substances in the ash.

                    (2 marks)

    (b)   What substances mentioned in the paragraph at the start of this exercise are found in the ash?

                    (1 mark)

    (c)   Suggest an explanation for the large amount of ash in the bullock carcass.

                    (2 marks)

3 Draw a bar chart to compare the substances present in dead leaves with those present in the bullock carcass.

                    (4 marks)

Hint   If you need help with drawing your bar chart look at Box 4 on page 17.

4 Give **one** place in dead leaves where lipids are found.

                    (1 mark)

5 Was the animal that produced the faeces a herbivore or a carnivore? Use information from the table to support your answer.

                    (2 marks)

Exercise 4.2 *continued*

We will now look at how different parts of the same organism differ from each other in the proportion of different substances they contain. The pie charts in Figure 4.3 show the composition of three different parts from the body of a bullock. The figures are percentages.

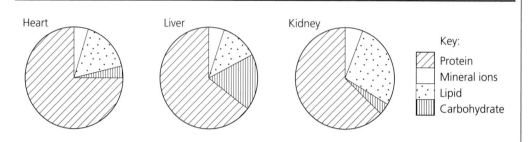

Figure 4.3 *The proportion of different substances in three different organs from a bullock*

6   (a)   Describe how you would show that a sample of heart contained protein.

(2 *marks*)

(b)   There is a high percentage of protein in the heart. Explain this in terms of the tissues present in the heart.

(1 *mark*)

7   (a)   Describe the main way in which the composition of the liver differs from the other two parts of the bullock shown in the pie charts.

(1 *mark*)

(b)   Suggest an explanation for this difference.

(2 *marks*)

Exercise 4.3 The composition of milk

In order to answer the questions in this exercise, you must

- make sure you understand how triglycerides are formed from glycerol and fatty acids
- know how saturated fatty acids differ from unsaturated fatty acids.

Milk provides a young mammal with all the nutrients it needs for its early growth and development. Milk contains proteins, carbohydrates and lipids. It also provides essential vitamins and mineral ions. The composition of milk, however, varies enormously. It varies from one species to another, from one individual to another, and even within a single individual. In this exercise we will look at one aspect of the composition of human milk – its lipid concentration.

The graph in Figure 4.4 shows how the concentration of triglyceride in human milk changes as lactation (the period of milk production) progresses.

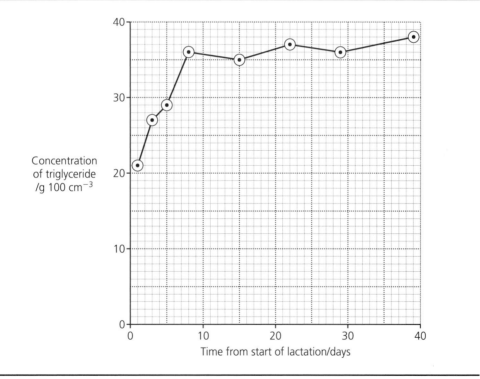

Figure 4.4 *The concentration of triglyceride in milk samples as lactation progresses*

1   In this graph, the amount of triglyceride is given as a concentration rather than as the total mass. Explain why.

(2 marks)

Exercise 4.3 *continued*

2   Describe how the concentration of triglyceride varies as lactation progresses.

(2 marks)

The composition of the milk that a woman produces depends to some extent on the food she eats. In a study of human breast milk, samples were collected from two groups of women. Those in one group were vegans. They only ate food obtained from plants. Those in the other group formed the control group. They ate food from both animal and plant sources. Table 4.3 shows the concentration of different fatty acids in milk samples from women in these two groups.

Table 4.3 The fatty acid composition of milk from groups of women on different diets

Fatty acid	Number of double bonds in hydrocarbon chain	Concentration of fatty acid in mg per g of lipid	
		Vegan group	Control group
Palmitic	0	166	276
Myristic	0	68	80
Stearic	0	52	108
Palmitoleic	1	12	36
Oleic	1	313	353
Linoleic	2	317	69
Linolenic	3	15	8

3   This investigation was conducted on a sample of eight women. There were four in the vegan group and four in the control group. Explain why the results of the investigation may not have been reliable.

(1 mark)

4   (a)   Lauric acid is a saturated fatty acid. Explain why it is described as saturated.

(1 mark)

        (b)   Linolenic acid is sometimes called a polyunsaturated fatty acid. Explain why.

(1 mark)

5   (a)   Look at the figures in Table 4.3. Describe the main differences between the occurrence of saturated and unsaturated fatty acids in the milk produced by these two groups of women.

(2 marks)

(b)   Suggest how variations in diet may explain these differences.

(1 mark)

In another study of human breast milk, samples obtained in 1953 and in 1987 were analysed and their fatty acid composition determined. The results from this study are given in Table 4.4.

Table 4.4 The fatty acid composition of breast milk samples obtained in 1953 and in 1987

Fatty acid	Number of double bonds in hydrocarbon chain	Concentration of fatty acid as percentage of total fatty acid content	
		1953	1987
Palmitic	0	23.2	22.5
Myristic	0	8.5	5.2
Stearic	0	6.9	8.7
Palmitoleic	1	3.0	4.1
Oleic	1	36.5	39.5
Linoleic	2	7.8	14.4
Linolenic	3	0	2.0

6   (a)   Plot the data shown in Table 4.4 as a suitable graph. Your graph should allow you to compare the total amounts of saturated and unsaturated fatty acids in the samples from the two years.

(3 marks)

(b)   Suggest an explanation for the differences between the two years.

(2 marks)

Exercise 4.4 Washing whiter than white

In order to answer the questions in this exercise, you must

- understand how investigations are planned and evaluated
- make sure you understand how triglycerides are formed from glycerol and fatty acids
- be able to explain how temperature and substrate concentration affect the rate of an enzyme-controlled reaction.

Since washing powders were first used for washing clothes, manufacturers have been trying to improve them by changing the ingredients. In the 1930s, enzymes were first added. This was not very successful because these enzymes needed very alkaline conditions and these conditions damaged the material from which the clothes were made. Since these early days, a lot of progress has been made and modern enzyme-based washing powders are very efficient at removing stains.

An investigation was carried out on the effect of temperature on stain removal by protein-digesting enzymes in washing powder.

- Identical pieces of cloth were stained with egg yolk.
- Each piece of cloth was put in a beaker containing 5 g of washing powder dissolved in 100 cm^3 of tap water.
- The beakers were then placed in water baths at different temperatures.
- The pieces of cloth were removed at regular intervals. The time taken for the egg stain to disappear was recorded.
- A control was set up at each temperature.

1 Which of the following would be the most suitable control for this investigation? Give a reason for your answer.
   A piece of cloth in a beaker containing

   A   100 cm^3 of tap water

   B   100 cm^3 of distilled water

   C   5 g of non enzyme-containing washing powder in 100 cm^3 tap water

   D   5 g of another enzyme-containing washing powder in 100 cm^3 tap water.

   (2 marks)

Exercise 4.4 *continued*

2   Give **two** reasons why the results from this investigation would be unreliable. Suggest how you could improve the investigation to overcome these problems.

(4 marks)

Lipolase is an enzyme which is added to washing powder. It hydrolyses triglycerides so it helps the washing powder to remove fat stains.

3   What products are formed by the action of lipolase on triglycerides?

(2 marks)

Figure 4.5 is a graph showing the effect of different concentrations of lipolase on the removal of lipid stains.

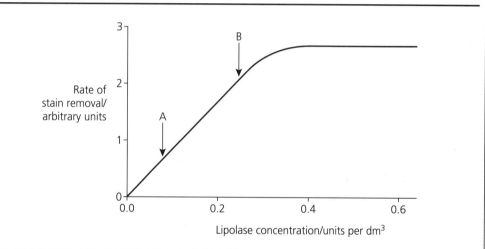

Figure 4.5 *The effect of different concentrations of lipolase on the rate of removal of lipid stains from a piece of cloth*

4   (a)   What is the evidence from Figure 4.5 that the rate of stain removal is limited by lipolase concentration between points **A** and **B**?

(1 mark)

   (b)   Explain how the rate of stain removal is limited by lipolase concentration between these two points.

(2 marks)

5 Explain why there is no further increase in the rate of stain removal when the lipolase concentration increases above 0.4 units per dm³.

(2 marks)

A new version of lipolase has been developed. It is called lipolase ultra. Figure 4.6 shows the effect of temperature on the relative efficiency of lipolase and lipolase ultra in removing stains.

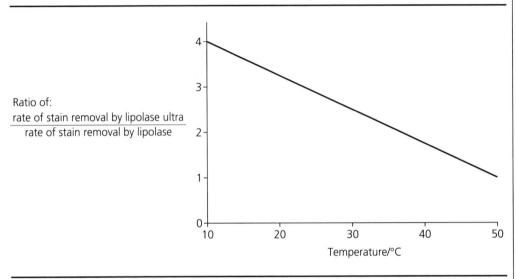

Figure 4.6 *A comparison of the relative efficiency of lipolase and lipolase ultra*

6 (a) At a temperature of 50 °C, the time taken to remove 1 g of fat stain by lipolase was 7 minutes. How long would it take lipolase ultra to remove 1 g of fat stain at the same temperature?

(1 mark)

   (b) In the USA and some Asian countries wash temperatures are usually between 10 °C and 20 °C. Use the graph to explain why lipolase ultra would be more suitable than lipolase for use in washing powders in these countries.

(1 mark)

Exercise 4.5 Enzyme shape and enzyme function

In order to answer the questions in this exercise, you must be able to explain how

- enzymes catalyse biochemical reactions
- proteins are formed by the folding of polypeptide chains.

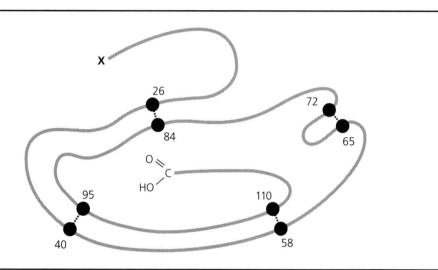

Figure 4.7 *A simple diagram showing the tertiary structure of a molecule of the enzyme RNAase. The large black dots represent cysteine, a sulphur-containing amino acid. The numbers represent positions along the polypeptide chain. Number 26, for example, is the 26th amino acid*

1 What chemical group is represented by the letter X on Figure 4.7?

(1 mark)

Hint In the rest of the questions in this exercise you will come across a number of substances which may be unfamiliar. Do not worry about them. It is what they do that is important, not their names.

2 All molecules of RNAase have the same tertiary structure.

(a) Explain what is meant by the tertiary structure of a protein.

(1 mark)

(b) Use information in Figure 4.7 to explain what causes all molecules of RNAase to have the same tertiary structure.

(3 marks)

Exercise 4.5 *continued*

In an investigation of the importance of the tertiary structure of enzymes, RNAase was treated with a mixture of mercaptoethanol and urea. This treatment breaks hydrogen bonds and disulphide bridges. The mercaptoethanol/urea mixture was then removed. As a result, the hydrogen bonds and disulphide bridges re-formed. After each treatment, the rate of reaction of the enzyme was measured. The results are shown in Figure 4.8.

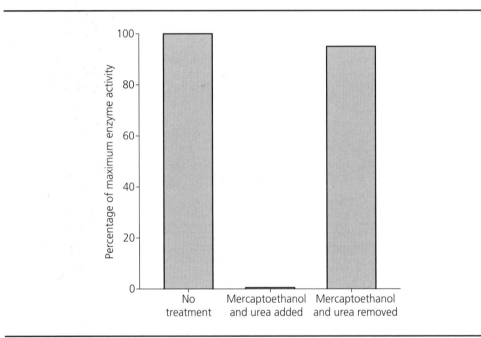

Figure 4.8 *The effect of different experimental treatments on the rate of reaction of RNAase*

3 When the rate of reaction was measured, a buffer solution was added to the enzyme–substrate mixture. Explain why it was necessary to add a buffer solution.

(2 marks)

4   (a)   How would you expect the mixture of mercaptoethanol and urea to affect the structure of RNAase molecules?

(1 mark)

(b)   Use your answer to question 4 (a) to explain the activity of the enzyme after mercaptoethanol and urea had been added.

(3 marks)

5 Suggest an explanation for the results obtained when the mercaptoethanol and urea were removed.

(1 mark)

Exercise 4.6 Getting through plasma membranes

In order to answer the questions in this exercise, you must be able to
  • describe the structure of a plasma membrane
  • explain how substances pass into and out of cells by simple diffusion, facilitated diffusion and active transport.

Many reactions take place in living cells. Cells respire and synthesize large molecules such as proteins, polysaccharides and nucleic acids. In order to carry out these reactions, they need to exchange substances with their surroundings. A cell must have, for example, a supply of oxygen, glucose and amino acids. At the same time it must get rid of waste products such as carbon dioxide and urea. The plasma membrane plays a very important part in regulating the passage of these and other substances into and out of a cell. Three processes are particularly important in allowing substances to pass through the plasma membrane. They are simple diffusion, facilitated diffusion and active transport. In this exercise we shall look at these processes in a little more detail.

1 Copy and complete Table 4.5. This table compares simple diffusion, facilitated diffusion and active transport. Put a tick (✓) in the box if the statement is true about the process or a cross (✗) if it is not true.

Table 4.5 Comparing simple diffusion, facilitated diffusion and active transport

Statement	Process		
	Simple diffusion	Facilitated diffusion	Active transport
Requires protein carrier molecules			
Requires energy in the form of ATP			
Transports substances from a low concentration to a high concentration			

(3 marks)

Hint Make sure that you complete Table 4.5 in the way you have been instructed in the question.

Exercise 4.6 *continued*

An artificial membrane was made. It consisted only of a bilayer of phospholipid molecules. In an investigation, the permeability of this artificial membrane was compared with the permeability of a plasma membrane from a cell. Figure 4.9 shows some of the results obtained from this investigation.

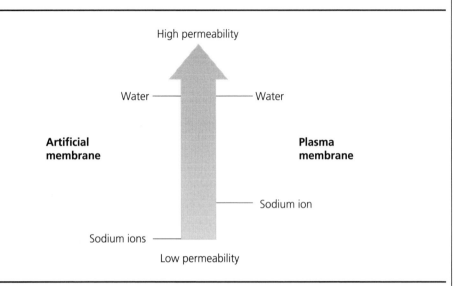

Figure 4.9 *A comparison of the permeability of an artificial membrane and a plasma membrane to water and sodium ions*

2 Explain why:

(a) both membranes allowed water molecules to pass through.

(2 marks)

(b) only the plasma membrane allowed sodium ions to pass through.

(2 marks)

Exercise 4.6 *continued*

In another investigation, the rate of uptake of glucose by red blood cells was measured. The cells were put in glucose solutions of different concentration. The graph in Figure 4.10 shows the results of this investigation. Curve **A** shows the results actually obtained. Curve **B** shows the results that would have been expected if glucose entered the cells by simple diffusion.

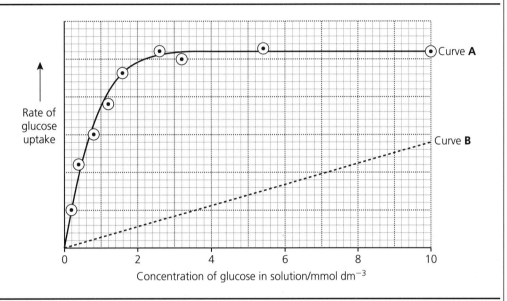

Figure 4.10 *The effect of glucose concentration on its rate of uptake by red blood cells*

Carrier proteins in the plasma membrane of a red blood cell help glucose to enter the cell by facilitated diffusion.

3   (a)   What limits the rate at which glucose molecules enter the cell below a glucose concentration of 2 mmol dm⁻³?

(1 mark)

       (b)   Give the evidence from Figure 4.10 that supports your answer.

(2 marks)

4 Explain why the curve flattens out above a glucose concentration of 2 mmol dm⁻³.

(2 marks)

5 Compare curve **B** with curve **A**. What is the advantage to a red blood cell of being able to take in glucose by facilitated diffusion rather than by simple diffusion?

(1 mark)

6 Xylose is a monosaccharide. The protein which transports glucose molecules across the plasma membrane into the red blood cell does not transport xylose. Use your knowledge of the tertiary structure of proteins to explain why.

(3 marks)

Tetracycline is a very useful antibiotic. It is used to treat a wide range of diseases caused by bacteria. Tetracycline works by inhibiting protein synthesis. It is thought that tetracycline is active against bacterial cells but not against animal cells because it can reach a concentration inside a bacterial cell of up to 30 times its concentration outside the cell.

7 What does the information above suggest about the way in which tetracycline gets into bacterial cells? Explain your answer.

(2 marks)

8 Suggest why tetracycline does not reach a high concentration inside an animal cell.

(1 mark)

Exercise 4.7 Plants, salts and water

In order to answer the questions in this exercise, you must be able to

- explain how substances pass into and out of cells by diffusion, osmosis and active transport.

Salinization is a word that we use to describe the build up of mineral salts in the soil. In many hot, dry areas of the world, the only way to grow crops is to irrigate them. The water used for irrigation contains dissolved salts. As this water evaporates, it leaves some of these salts behind and they build up in the surface layers of the soil. In these semi-desert regions the water table often lies far underground. The water added during irrigation causes this water table to rise and it brings with it more dissolved salts to accumulate in the surface layers. If these processes continue for too long, the concentration of salts becomes too high and many species of plants are no longer able to survive.

In this investigation, we will look at how salinization affects the growth of plants and the yield of crops. Before we do this, we will look at how plants absorb water and mineral ions from the soil. Look at Table 4.6.

Table 4.6 The concentration of different ions in the soil and in the root tissue of pea plants

Ion	Concentration/mmol dm^{-3}	
	in soil	in root tissue
K^+	1.0	75.0
Na^+	1.0	8.0
Mg^{2+}	0.3	1.5
NO_3^-	2.0	27.0
SO_4^{2-}	0.3	9.5

1   (a)   What is the evidence from the figures in Table 4.6 that pea plants take up ions from the soil by active transport?

(1 mark)

(b)   Would you expect the water potential of the pea root tissue to be less negative or more negative than the water potential of the surrounding soil? Give a reason for your answer.

(1 mark)

(c)   Use your answer to question 1 (b) to explain how water normally enters the root from the soil.

(1 mark)

Exercise 4.7 *continued*

Hint   Do not be put off because this question appears to be asking you about plants! All you are really required to do is to apply your knowledge of how substances get in and out of cells to material which you are unlikely to have encountered before.

The graph in Figure 4.11 shows the effect of oxygen concentration on the rate of uptake of potassium ions by pea roots.

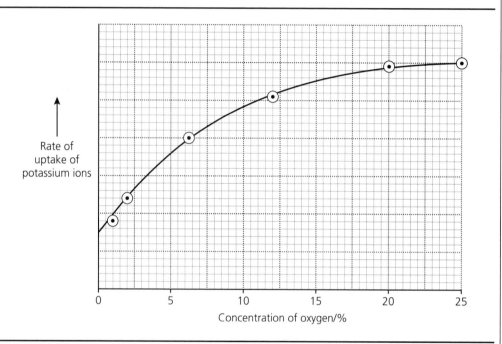

Figure 4.11 *The effect of oxygen concentration on the rate of uptake of potassium ions by pea roots*

2   (a)   Using Figure 4.11, describe the effect of increasing oxygen concentration on the rate of uptake of potassium ions.

(2 marks)

(b)   Explain the shape of the curve between oxygen concentrations of 5 per cent and 20 per cent. In your answer you should refer to

• respiration
• active transport.

(3 marks)

The answers to questions 1 and 2 should tell you about the way in which a plant such as a pea plant absorbs ions and water when it is growing in normal soil. We will now consider what happens when it is growing in soil with a high concentration of salts.

Exercise 4.7 *continued*

3   (a)   How will a high concentration of salts in the soil affect the soil water potential?

(1 mark)

(b)   Explain how the soil water potential resulting from a high concentration of salts would affect the rate of water uptake by a plant.

(2 marks)

Some plants have adaptations that enable them to grow in soils which have a high concentration of salts. One of these is a species of grass called *Cynodon dactylon*. The effect of salt concentration on the cells of this grass was investigated. Table 4.7 shows some of the results of an investigation.

Table 4.7 The effect of salt concentration on water potential of root cells and on the rate of photosynthesis of *Cynodon dactylon*

Salt concentration of the soil/mol dm^{-3}	Water potential of root cells/kPa	Rate of photosynthesis/arbitrary units
20	−355	21.4
100	−395	21.5
180	−895	29.4
260	−1300	24.7
340	−1420	29.4

4   It has been suggested that this grass survives in soils containing a high concentration of salts because it absorbs more ions into its cells in these conditions.

(a)   Explain how an increase in the concentration of ions in the root cells might help a plant to gain enough water and survive in soils which have a high salt concentration.

(2 marks)

(b)   What is the evidence from Table 4.7 that the cells of this grass do have a higher concentration of ions in soils where the salt concentration is high.

(1 mark)

Exercise 4.7 *continued*

5   Although this grass can survive in soils which have a high concentration of salts, it does not grow well. It has been suggested that this is because the high concentration of ions in its cells slow down the rate of photosynthesis. Do the results in Table 4.7 support this idea? Explain your answer.

(1 mark)

Hint   You do not need to know anything about photosynthesis to answer this question other than that it is the way in which plants produce many of the substances they require for growth.

CHAPTER

five | Genes

 Exercise 5.1 Cells, chromosomes and DNA

In order to answer the questions in this exercise, you must be able to explain

- what happens to the number of chromosomes in a cell as a result of mitosis, meiosis and fertilization
- how the mass of DNA in a cell changes during the cell cycle.

Different species of organisms have different numbers of chromosomes in their body cells (somatic cells). Table 5.1 shows the number of chromosomes in the body cells of some different organisms.

Table 5.1 The number of chromosomes in the body cells of different organisms

Species	Number of chromosomes in the nucleus of a body cell
Human	46
Horse	64
Donkey	62
Cabbage	18
Wheat	42
Yeast	34

1 Explain why all the numbers given in Table 5.1 are even numbers.

(1 mark)

2 A mule is the offspring of a horse and a donkey. The horse is the female parent and the donkey is the male parent.

(a)   How many chromosomes are there in a body cell from a mule? Explain how you arrived at your answer.

(3 marks)

(b)   Suggest why mules are sterile and cannot produce offspring.

(2 marks)

Mosses are small plants which have complex life cycles. In this life cycle, a spore-producing plant alternates with a gamete-producing plant. Figure 5.1A shows a moss. You can see the spore-producing plant growing on the gamete-producing plant. Figure 5.1B shows the complete life cycle of a moss. Each of the boxes on this diagram shows the number of chromosomes in one of the cells from the relevant stage.

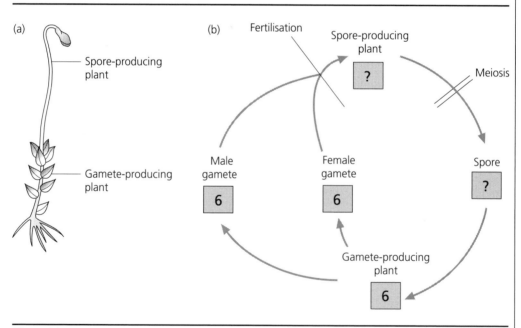

Figure 5.1A *shows a moss with spore-producing and gamete-producing plants.* **Figure 5.1B** *shows the life cycle of one particular species of moss*

Exercise 5.1 *continued*

3 How many chromosomes are there in

   (a)   a cell from a spore-producing plant?

   (b)   a spore?

   (2 marks)

4 By what type of cell division does the gamete-producing plant produce male gametes?

   (1 mark)

Hint   The life cycle of a moss may be unfamiliar to you but all the information you will need to answer questions 3 and 4 is provided in the question.

The bar chart in Figure 5.2 shows some of the results of an investigation carried out over 50 years ago. It shows the mass of DNA present in the nuclei of cells from different species of animal.

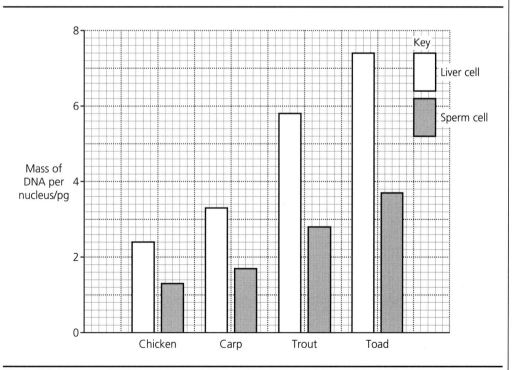

Figure 5.2 *The mass of DNA present in nuclei from different cells in four species of animal*

Exercise 5.1 *continued*

5   (a)   What is the approximate ratio of the mass of DNA in the nucleus of a liver cell to the mass of DNA in the nucleus of a sperm cell?

(1 mark)

(b)   Explain the link between this ratio and the chromosomes in these cells.

(1 mark)

6   Giving an explanation for your answer in each case, estimate the mass of DNA you would expect to find in the nucleus of

(a)   a cell from the lining of the small intestine of a trout.

(2 marks)

(b)   a toad egg cell which had just been fertilized.

(2 marks)

Exercise 5.2 Chromosomes and cell division

In order to answer the questions in this exercise, you must be able to

- describe the process of mitosis.

Figure 5.3 shows a cell which is undergoing mitosis. This cell was observed carefully and the following measurements taken:

- the distance between the poles of the cell (**X**)
- the distance between the sister chromatids (**Y**)
- the distance between the sister chromatids and the poles of the cell (**Z**).

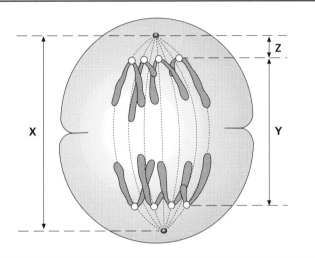

Figure 5.3 *A cell undergoing mitosis*

Exercise 5.2 *continued*

These measurements were then plotted on the graph shown in Figure 5.4.

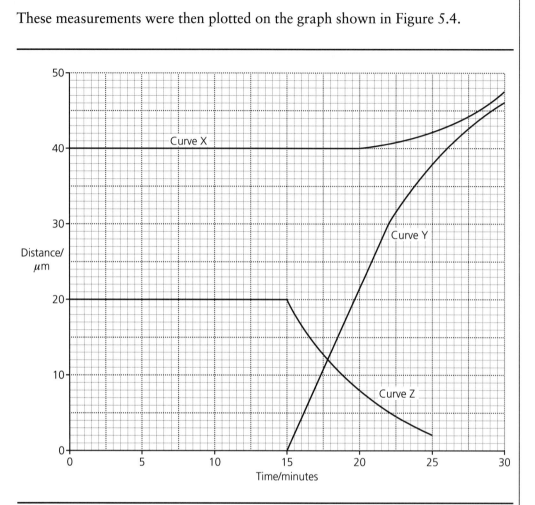

Figure 5.4 *Graph showing the movement of chromosomes during mitosis*

Hint Although you will need to know the names of the stages of mitosis, it is more important to be able to explain what happens during cell division. The questions in this exercise will help you with this.

1 (a) Describe what happens to sister chromatids during anaphase of mitosis.

(2 marks)

(b) Use curve **Z** on the graph and your answer to question 1 (a) to estimate how long anaphase took in this cell. Explain how you arrived at your answer.

(2 marks)

Exercise 5.2 *continued*

2 The sister chromatids move apart from each other during mitosis. Use curve **Y** to calculate the maximum rate at which the sister chromatids moved apart. Give your answer in mm minute^{-1} and show your working.

(2 marks)

Hint Questions are often set on calculating rate. Make sure that you can do this.

3 Explain the evidence from the graph that curve **X** represents the distance between the poles of the cell.

(2 marks)

4 Some drugs which are used in the treatment of cancer prevent the formation of spindle fibres during mitosis.

   (a)   Explain how these drugs would prevent the formation of new cancer cells.

(2 marks)

   (b)   How would curve **Z** have differed if the cell had been incubated with one of these drugs during mitosis?

(1 mark)

If you look at a section through the tip of a root with a microscope, you will see that many of the cells are dividing by mitosis. In an investigation, numbers of cells in different stages of mitosis in a root tip were counted. The results are shown in Table 5.2.

Table 5.2 The number of cells in different stages of mitosis in a root tip

Stage	Number of cells
Prophase	210
Metaphase	30
Anaphase	12
Telophase	48

Exercise 5.2 *continued*

5   (a)   What do the data in Table 5.2 suggest about the lengths of the different stages of mitosis? Give the reason for your answer.

(2 marks)

           (b)   In this root tip, mitosis takes about two hours. Use the data in Table 5.2 to estimate the time a cell is in metaphase. Show your working.

(2 marks)

Exercise 5.3 Dividing cells

In order to answer the questions in this exercise, you must be able to

- explain how the mass of DNA in a cell changes during the cell cycle
- describe the structure of DNA and explain how DNA replicates.

Cells pass through a cycle of division and growth called the cell cycle. During mitosis, the nucleus of the parent cell divides to give two daughter nuclei. These daughter nuclei are genetically identical. During the long period of interphase which follows, many things happen in the cells which are formed from this division. These include the synthesis of different proteins and DNA replication. At the end of interphase, the cells divide again. We will look at the cells which line the small intestine.

1   The cells lining the small intestine have a very high rate of mitosis. Explain why you would expect a high rate of mitosis in these cells.

(2 marks)

The small intestine is lined with thousands of small finger-like projections called villi. Each villus is covered with a layer of cells which separate it from the cavity or lumen of the intestine. This layer of lining cells is called the epithelium. Figure 5.5 is a diagram showing these features.

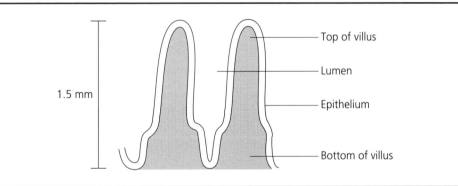

Figure 5.5 *Diagram showing the villi which line the small intestine*

In an investigation of mitosis in the epithelial cells of the small intestine, radioactive thymine was used. Thymine is a base found in nucleic acids. It was radioactively labelled and injected into the blood of a mammal. Some of this radioactive thymine was absorbed from the blood by the epithelial cells. It was eventually used in these cells to produce new DNA.

Hint   If you are not sure about why radioactively labelled substances are used in biological investigations, look back at Exercise 4.1 on page 50.

Exercise 5.3 *continued*

2 Explain how semi-conservative replication results in the DNA in these epithelial cells being labelled with radioactive thymine.

(2 marks)

3 Explain why radioactive thymine was used to label the DNA in these cells rather than

    (a)   radioactive uracil.

    (b)   radioactive cytosine.

(2 marks)

4 Explain why the daughter cells produced when the epithelial cell divides by mitosis both contain radioactive thymine.

(3 marks)

In this investigation, the time when radioactive DNA was first found in the epithelial cells at different distances along a villus was recorded. The results are shown in Figure 5.6.

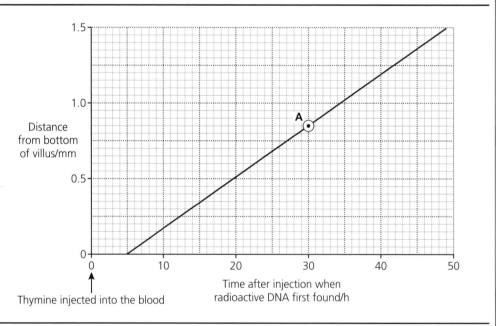

Figure 5.6 *Graph showing the time when radioactive DNA was first found in epithelial cells at different distances along a villus*

Exercise 5.3 *continued*

5 Look at point **A** on the graph. What does it tell you about the location of cells containing radioactive DNA?

(2 marks)

6 No radioactive DNA was found in any of the cells for 5 hours after the thymine was injected. Suggest **two** reasons why no radioactive DNA was not found in any of the cells.

(2 marks)

The scientists who carried out the investigation came to the conclusion that new cells are produced at the bottom of the villus. They also concluded that these cells are pushed upwards as more new cells are produced underneath them.

7 Give the evidence from the graph that

(a) new cells are produced at the bottom of the villus.

(1 mark)

(b) these cells are pushed upwards.

(1 mark)

Exercise 5.4 DNA, cells and organisms

In order to answer the questions in this exercise, you must

- know how a centrifuge separates different substances
- be able to describe the structure of DNA and understand how its bases pair.

The diagram in Figure 5.7 shows *Euglena viridis*. It is a single-celled organism.

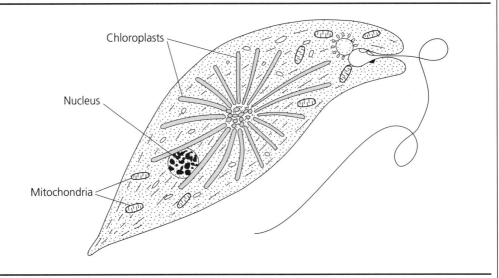

Figure 5.7 Some of the organelles found inside a cell of Euglena viridis

Some *Euglena viridis* cells were crushed and the DNA extracted. DNA was also extracted from isolated mitochondria and from isolated chloroplasts. These three samples of DNA were spun separately in a centrifuge. Their positions in the centrifuge tubes after spinning were recorded. The results are shown in the graph (Figure 5.8).

Exercise 5.4 *continued*

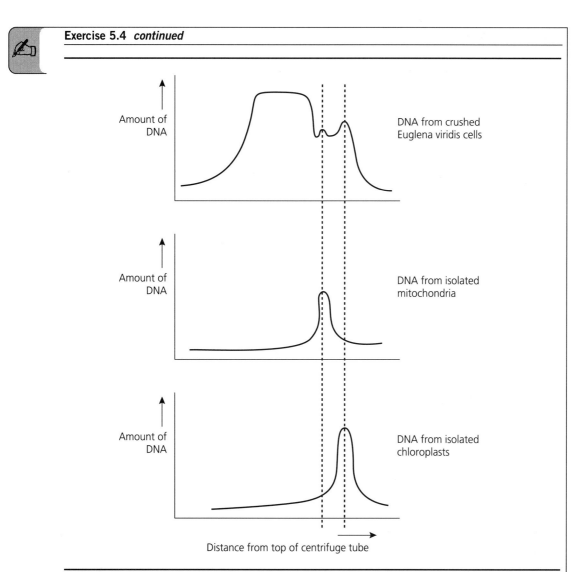

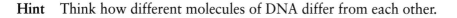

Figure 5.8 *Graph showing the location of DNA obtained from centrifuging samples from three different sources*

1 Explain the shape of the curve showing the DNA from crushed *Euglena viridis* cells.

(2 marks)

2 (a) When different substances are centrifuged, they finish in different positions in the centrifuge tube. What causes them to be in different positions?

(1 mark)

   (b) What features of DNA molecules result in different samples of DNA being in different positions after it has been centrifuged?

(2 marks)

Hint Think how different molecules of DNA differ from each other.

Exercise 5.4 *continued*

3   Use information from Figure 5.8 to suggest the number of peaks you would find if DNA from each of the following sources was centrifuged:

   (a)   human liver cells.

   (b)   cells from a plant root.

(2 marks)

We can produce a hybrid DNA molecule by taking one strand from one organism and joining it to the complementary strand taken from another organism. In the hybrid DNA molecule some base pairs are complementary but some may not be. Where the base pairs are complementary, bonds form which hold the two strands together. Look at Figure 5.9.

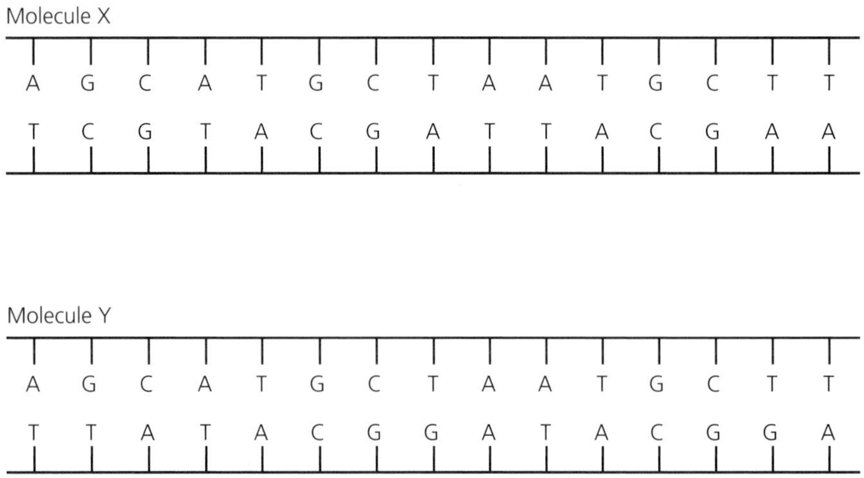

Molecule X

A G C A T G C T A A T G C T T
T C G T A C G A T T A C G A A

Molecule Y

A G C A T G C T A A T G C T T
T T A T A C G G A T A C G G A

Figure 5.9 *A simplified diagram showing the structure of two hybrid DNA molecules*

4   (a)   Name the type of bond which joins the complementary base pairs together.

(1 mark)

   (b)   Between how many base pairs in molecule **Y** would you expect these bonds to be formed?

(1 mark)

Exercise 5.4 *continued*

(c)   One of the hybrid DNA molecules in Figure 5.9 was formed by joining strands from different species. Which molecule was this? Give a reason for your answer.

(1 mark)

When a molecule of DNA is heated, it splits into its two strands as the bonds holding these strands together are broken. The strands separate as a result. The more bonds which join bases together, the higher the temperature required to separate the strands. Table 5.3 shows some of the results obtained when using a technique based on this principle to study how closely humans are related to other species of primate.

Table 5.3 The differences between the bases present in human DNA and the bases present in the DNA of other species of primate

Hybrid DNA molecule formed between a strand of human DNA and a strand of DNA from the following:	Percentage of bases present in human DNA which differ from bases present in the DNA of the species tested
human	
chimpanzee	2.4
gibbon	5.3
green monkey	9.5
capuchin monkey	15.8
galago	42.0

5   Complete Table 5.3 by giving the percentage of bases which you would expect to be different in a hybrid DNA molecule formed between two strands of human DNA.

(1 mark)

6   Which of the hybrid DNA molecules in the table would you expect to be separated into its component strands at the lowest temperature? Give the reason for your answer.

(2 marks)

7   To which primate in the table is a human most closely related? Explain your answer.

(2 marks)

Exercise 5.5 Bases and base sequences

In order to answer the questions in this exercise, you must be able to
- describe the structure of DNA and understand how its bases pair
- describe the structure of mRNA and understand how it is formed during transcription.

1   Copy and complete Table 5.4 to compare the structure of a DNA molecule from a bacterial cell and a DNA molecule from a human cell. Use a tick (✓) if the statement is correct or a cross (✗) if it is not correct.

(2 marks)

Table 5.4 A comparison of DNA from a bacterial cell and DNA from a human cell

Statement	DNA from a bacterial cell	DNA from a human cell
DNA molecule has its ends joined together to form a loop		
DNA molecule consists of two strands joined by hydrogen bonds		
DNA has two complementary base pairs, adenine and thymine and cytosine and guanine		

Samples of DNA were extracted from a number of different species of bacteria. These samples were analysed. Table 5.5 shows the percentage of different bases in the DNA from each species.

Table 5.5 The percentage of different bases in the DNA from different species of bacteria

Species	Base / %			
	Adenine (A)	Guanine (G)	Cytosine (C)	Thymine (T)
Escherichia coli	24.7	26.0	25.7	23.6
Staphylococcus aureus	30.8	21.0	19.0	29.2
Clostridium perfringens	36.9	14.0	12.8	36.3
Sarcina lutea	13.4	37.1	37.1	12.4

2   (a)   Calculate the ratio of adenine to thymine for each of the four species of bacteria.

(1 mark)

    (b)   Use your knowledge of the structure of DNA to predict the ratio of adenine to thymine in a DNA molecule.

(1 mark)

    (c)   Suggest an explanation for the slight difference between the actual ratios of adenine to thymine found in this investigation and the predicted ratio.

(1 mark)

Hint   You have to calculate a number of ratios in this question. Always express a ratio in its simplest form, preferably as something to 1.

3   The amount of guanine in the bacterium *Brucella abortis* is 29 per cent. Estimate the percentage of the other following bases found in this species of bacterium:

    (a)   cytosine.

    (b)   adenine.

    In each instance, explain how you arrived at your answers.

(3 marks)

4   Calculate the ratio of adenine to guanine in each of the different species of bacteria in Table 5.5.

    (a)   What do you notice about these ratios compared to the ratios of adenine to thymine?

(1 mark)

    (a)   Suggest an explanation for your answer to question 4 (a).

(2 marks)

Exercise 5.5 *continued*

Organs in most animals get larger because the number of cells in the organ increases. These cells go through a cycle of growth and division. Some insect organs grow in a different way. Their cells simply get larger and larger. These large cells contain giant chromosomes. Giant chromosomes are very large and are made up from many molecules of DNA which lie parallel to each other and are clearly visible at all stages of the cell cycle. Figure 5.10 shows part of a giant chromosome.

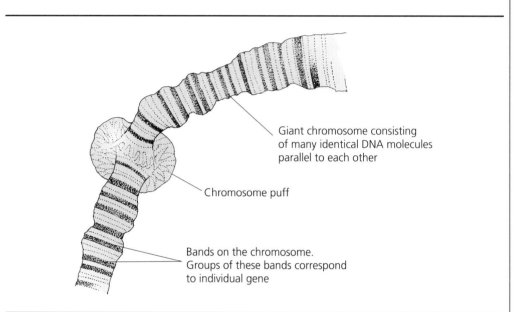

Giant chromosome consisting of many identical DNA molecules parallel to each other

Chromosome puff

Bands on the chromosome. Groups of these bands correspond to individual gene

Figure 5.10 Part of a giant chromosome from a cell in a salivary gland from a fly

The chromosome puffs shown in the diagram are regions of the chromosome where a gene is active and transcription is taking place. Molecules of mRNA from chromosome puffs at two different places on a giant chromosome were analysed. The results are shown in Table 5.6.

Table 5.6 The percentage of different bases in mRNA from chromosome puffs at different places on a giant chromosome

Location of chromosome puff	Base / %			
	Adenine	Guanine	Cytosine	X
Middle of chromosome	38.0	20.5	24.5	
End of chromosome	31.2	22.0	26.4	

Exercise 5.5 *continued*

5   (a)   Name base **X**.

(1 mark)

(b)   Calculate the percentage of base **X** in the two samples of mRNA.

(1 mark)

6   Explain why the percentage of the bases in the two mRNA samples differ from each other.

(2 marks)

Exercise 5.6 RNA and protein synthesis

In order to answer the questions in this exercise, you must

- know that there are different types of RNA
- be able to explain the part played by the different types of RNA in the synthesis of proteins.

The total mass of RNA in a single mouse cell is so small it has to be measured in picograms and a picogram is one million millionth of a gram or 1×10^{-12} grams. There are approximately 26 pg of RNA in a mouse cell. Although this may seem to be a tiny amount, it is around 50 000 million nucleotides!

There are a number of different types of RNA. These include the RNA present in the nucleus and the RNA present in the cytoplasm. Table 5.7 shows the amounts of these different types of RNA in a mouse cell.

Table 5.7 The percentage of different types of RNA in a typical mouse cell

Type of RNA	Percentage of total RNA
RNA present in nucleus	11
rRNA	71
mRNA	3
tRNA	15

1   Calculate the total mass of tRNA in a typical mouse cell.

(1 mark)

2   Name the organelle in this cell in which you would expect to find the greatest mass of RNA. Give the reason for your answer.

(2 marks)

3   In an animal cell such as this mouse cell there are between 10 000 and 20 000 different sorts of mRNA. There are many fewer sorts of tRNA.

(a)   Explain why a very large number of different sorts of mRNA are necessary for a mouse cell to carry out its functions.

(2 marks)

Exercise 5.6 *continued*

    (b)   Use your knowledge of the genetic code to estimate the number of different sorts of tRNA which would be found in the cell. Explain how you arrived at your answer.

(3 marks)

Insects such as flies lay eggs. These eggs hatch into larvae which feed and grow rapidly. When a larva is fully grown, it moults and becomes a pupa. Inside the pupa, many changes take place. The organs of the larva break down and the new organs of an adult fly are formed. Eventually the pupal 'skin' splits and a new adult fly emerges. In an investigation of the changes which take place inside a pupa, the total amount of RNA was measured at different times. The results are shown in Figure 5.11.

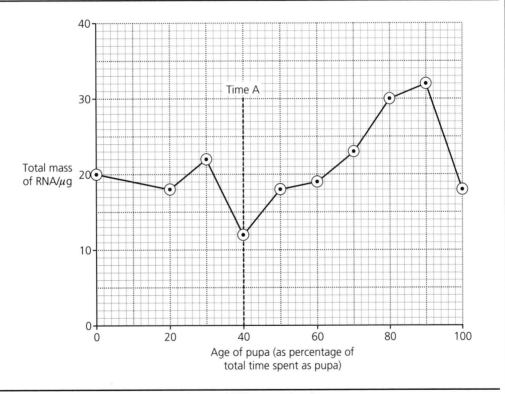

Figure 5.11 *The effect of age on the total mass of RNA present in a fly pupa*

Exercise 5.6 *continued*

4   (a)   Explain why the points on the graph have been joined to each other with straight lines.

(1 mark)

(b)   The length of time a fly spends as a pupa depends on temperature. Suggest why the investigators gave the age of the pupa as the percentage of the total amount of time spent as a pupa.

(2 marks)

5   (a)   Describe how the mass of RNA changes during the period that the fly spends as a pupa.

(2 marks)

(b)   Use your knowledge of the function of RNA to explain its change in mass after time **A**.

(3 marks)

CHAPTER six

Gas exchange and transport

Exercise 6.1 Breathing in and breathing out

In order to answer the questions in this exercise, you must be able to

- describe the path taken by inhaled and exhaled air into and out of the lungs.

The amounts of different gases in samples of inhaled air and exhaled air can be measured using the simple piece of apparatus shown in Figure 6.1.

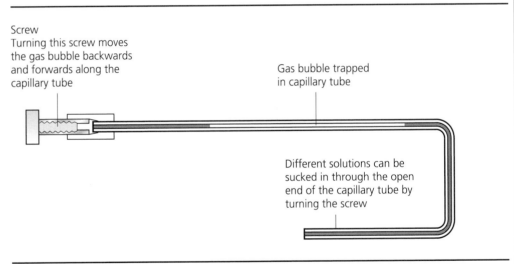

Screw
Turning this screw moves the gas bubble backwards and forwards along the capillary tube

Gas bubble trapped in capillary tube

Different solutions can be sucked in through the open end of the capillary tube by turning the screw

Figure 6.1 *A capillary J-tube used to analyse a sample of air*

Turning the screw at the end of the capillary tube moves an air bubble backwards or forwards along the tube. In this way, different solutions can be sucked into the open end of the tube. These solutions wet the inside of the tube and absorb different gases from the trapped air bubble. The solutions used and the measurements made are summarized in Table 6.1. The solutions are used in the order shown, with alkaline pyrogallol used only after the potassium hydroxide has removed all the carbon dioxide from the air bubble.

Exercise 6.1 *continued*

Table 6.1 Analysing a gas bubble trapped in a capillary J-tube

Solution sucked into capillary tube	Gas absorbed	Length of gas bubble after treatment
Initial length of air bubble		A
Potassium hydroxide	Carbon dioxide	B
Alkaline pyrogallol	Oxygen	C

1   Use the letters **A**, **B** and/or **C** to write a simple formula which would allow you to calculate

(a)   the amount of carbon dioxide in the original sample.

(1 mark)

(b)   the percentage of carbon dioxide in the initial sample.

(1 mark)

2   (a)   Alkaline pyrogallol contains potassium hydroxide solution. Explain why alkaline pyrogallol is only used after all the carbon dioxide has been absorbed.

(1 mark)

(b)   Use the letters **A**, **B** and/or **C** to write a simple formula which represents the amount of oxygen in the original sample.

(1 mark)

3   The apparatus was left submerged in a bowl of water at 20 °C between each set of readings. Explain why.

(2 marks)

Exercise 6.1 *continued*

We will look at how a sample of exhaled air may be obtained for analysis. Air enters the lungs through the nose and mouth. It then goes down the trachea and through the bronchi into the alveoli.

4 (a) When you breathe out, the air you first exhale is very similar in composition to inhaled air. Explain why.

(2 marks)

   (b) If you continue to breathe out, the air at the end of the breath has been in the alveoli of the lungs. Suggest how you could obtain a sample of this air for analysis with a capillary J-tube.

(2 marks)

Table 6.2 shows some results obtained with this apparatus.

Table 6.2 The percentage of gases present in samples of inhaled and exhaled air

Gas	Percentage of gas present in a sample of	
	inhaled air	exhaled air
Oxygen	21	16
Carbon dioxide	0	4
Nitrogen and other gases	79	80

5 (a) The actual figure for the percentage of carbon dioxide present in the atmosphere is 0.04. Explain why the figure given in Table 6.2 is 0.

(2 marks)

   (b) Suppose the sample of air had been taken from a complete breath rather than just from the air in the alveoli. Would you have expected the percentage of oxygen to have been the same, more, or less than that shown in the third column of the table? Explain your answer.

(3 marks)

Exercise 6.2 Gas exchange in the lungs

In order to answer the questions in this exercise, you must

- be able to explain how respiratory gases are exchanged in the lungs
- understand the part played by diffusion in gas exchange.

The partial pressure of oxygen (pO_2) is a measure of the amount of oxygen present in a mixture of gases. It is usually given in kilopascals (kPa). These are units of pressure. Figure 6.2 shows how the partial pressure of oxygen changes as blood passes through the pulmonary artery, the pulmonary capillaries and the pulmonary vein. The numbers on the x-axis show the length of time the blood has been in the pulmonary capillaries.

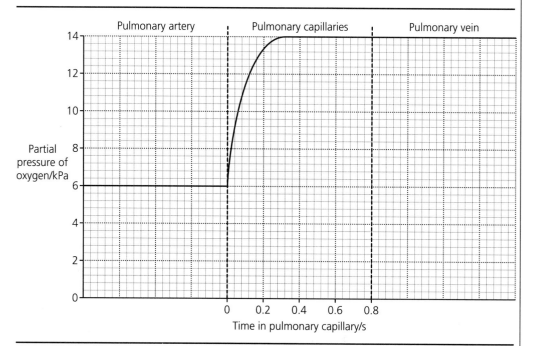

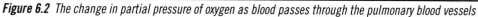

Figure 6.2 *The change in partial pressure of oxygen as blood passes through the pulmonary blood vessels*

1   (a)   Explain what causes the change in the partial pressure of oxygen
          in the blood as it passes through the pulmonary capillaries.

(2 marks)

    (b)   Use the information in Figure 6.2 to estimate the partial pressure
          of oxygen in the alveoli of the lungs. Explain your answer.

(3 marks)

    (c)   When a person takes vigorous exercise, the blood in the pulmonary
          capillaries flows faster. Use this information to explain why the
          partial pressure of oxygen entering the pulmonary vein is lower
          during vigorous exercise than when a person is resting.

(2 marks)

Exercise 6.2 *continued*

2   (a)   Copy the axes shown in Figure 6.3. Sketch a curve on these axes to show how you would expect the partial pressure of carbon dioxide to change as the blood flows from the pulmonary artery through the pulmonary capillaries to the pulmonary vein.

(2 marks)

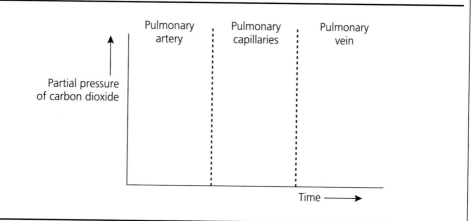

Figure 6.3 *You will complete this figure to show how you would expect the partial pressure of carbon dioxide to change in the blood as it passes through the pulmonary blood vessels*

   (b)   Emphysema is a condition in which the walls between the alveoli break down and enlarge the air spaces in the lungs. The blood of a person with emphysema has a higher concentration of carbon dioxide than the blood of a healthy person. Explain why.

(2 marks)

Hospitals sometimes carry out tests to see how well the lungs are working. One way in which they do this is to measure the gas transfer factor. This is done by measuring how much carbon monoxide is taken up from a single breath of air containing 0.3 per cent carbon monoxide.

3   (a)   By what process would carbon monoxide pass from the air in the alveoli to the blood in the pulmonary capillaries?

(1 mark)

   (b)   Suggest why carbon monoxide, not oxygen, is used for this test.

(2 marks)

4 In a person with interstitial lung disease, the alveolar walls become thicker. Explain why the gas transfer factor would be low in a person with interstitial lung disease.

(1 mark)

Exercise 6.3 Breathing and exercise

In order to answer the questions in this exercise, you must be able to

- explain how breathing is affected by exercise.

An investigation was carried out into the effects of exercise on breathing. An athlete pedalled an exercise bicycle on which the amount of work could be varied. At each exercise rate, measurements were made of breathing rate and tidal volume. The results of the investigation are shown in Table 6.3.

Table 6.3 The effect of exercise rate on breathing rate and tidal volume

Exercise rate/watts	Breathing rate/ breaths minute^{-1}	Tidal volume/dm^3
0	14.0	0.74
30	15.1	1.43
90	14.5	2.34
120	15.1	2.76
150	14.8	3.25
180	21.5	3.21

1 Calculate the total volume of oxygen taken into the lungs in one minute at an exercise rate of 30 watts. Assume that the proportion of oxygen present in the air is 21%. Show your working.

(3 marks)

2 What conclusions can you draw about the way in which breathing rate and tidal volume change with exercise rate?

(2 marks)

3 Describe how the medulla increases breathing rate.

(3 marks)

4 The athlete in this investigation had a vital capacity of 5.6 dm^3.

  (a)   Explain what is meant by *vital capacity*.

                                                    (2 marks)

  (b)   Showing your working in each case, calculate the tidal volume as a
        percentage of the vital capacity when the exercise rate was

        (i) 0 watts.
        (ii) 180 watts.

                                                    (2 marks)

5 Use the information in this question to write a short paragraph about how the
  pattern of breathing changes with exercise rate. In your account refer to each
  of the following:

  • breathing rate
  • tidal volume
  • vital capacity.

                                                    (3 marks)

Hint   In question 5 you are asked to put together a number of ideas. When you
           have answered this question, read it through carefully and make sure that
           your account is clear and easy to follow.

Exercise 6.4 Size, surface area and volume

In order to answer the questions in this exercise, you must be able to

- describe the relationship between size and the surface area to volume ratio of an organism
- make use of this relationship in explaining gas exchange and heat loss.

As an organism increases in size, so its surface area and volume also increase. What is not so obvious is how the ratio of its surface area to volume will change. Many attempts have been made to find out by measuring the surface area and volume of different animals.

1   Explain why there are likely to be errors in finding an animal's surface area by

   (a)   removing its skin and measuring the surface area of this.

(1 mark)

   (b)   comparing the animal with a series of cylinders and cones and calculating the surface area from these.

(1 mark)

2   It is much easier to measure the mass of an animal than it is to measure its volume. As an animal grows, will its mass be directly proportional to its volume? Explain your answer.

(2 marks)

The easiest way to investigate changes in the ratio of surface area to volume is to consider cubes which differ in size.

3 The surface area to volume ratio may be given as the result of dividing surface area by volume. Copy and complete Table 6.4 to show the surface area to volume ratio of different-sized cubes. One row has been completed to help you.

(2 marks)

Table 6.4 The surface area to volume ratio of cubes with different length sides. For clarity, the units have been left out of this table

Side length	Surface area	Volume	Surface area / Volume
1			
2			
3	54	27	2.0
4			
5			
6			

4 Plot the figures in the table as a suitable graph to show how the surface area to volume ratio of a cube varies with side length.

(4 marks)

Hint It is worth learning this graph so that you can sketch it if you need to. You will find it very useful in answering general questions on gas exchange and heat loss.

5 Which of the following statements are true and which are not true about the information in the table.

    A   The larger the cube, the greater its surface area.

    B   The smaller the cube the smaller its surface area in relation to its volume.

    C   As cubes get larger, their volume increases faster than their surface area.

    D   An increase in the size of a cube is associated with a decrease in the surface area to volume ratio.

(2 marks)

Exercise 6.4 *continued*

The eggs of many fish are only about 1 mm in diameter. When they hatch, they produce very small larvae. Gas exchange in these small larvae takes place only through the skin. As the fish increase in size, they can no longer obtain all the oxygen they require through the skin.

6 Newly hatched fish larvae can obtain all the oxygen they require by diffusion through the skin but older fish are unable to do so. Use your graph to explain why.

(3 marks)

The ratio of surface area to volume is also important when considering heat loss in mammals and birds. The graph in Figure 6.4 shows the percentage of daylight hours spent feeding in three British species of bird.

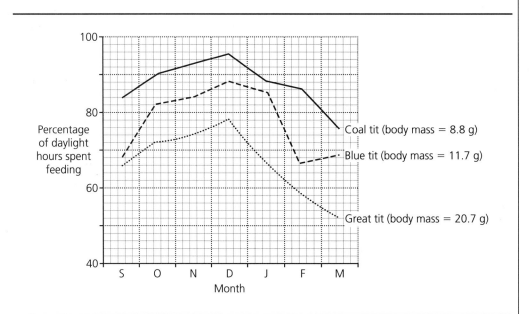

Figure 6.4 *The percentage of daylight hours spent feeding between September and March by coal tits, blue tits and great tits*

7 The body temperature of a bird is approximately 40 °C. Explain what causes the body temperature of a bird to be higher than that of its surroundings.

(2 marks)

8 Describe and explain the relationship between body mass and the percentage of time spent feeding. Use the graph that you drew earlier to help you to answer this question.

(3 marks)

Exercise 6.5 Heart beat

In order to answer the questions in this exercise, you must be able to

• explain the way in which a heart beat is initiated and coordinated.

The heart is made of muscle. This muscle is unlike other muscle in the body because it does not require a nerve impulse to make it contract. It beats on its own. A heart beat starts with an electrical signal from an area of muscle in the wall of the right atrium. This area of muscle is the sinoatrial node (SAN). A wave of electrical activity spreads from here and coordinates the heart beat. Figure 6.5 is a drawing of the human heart. The magnification of this drawing is × 0.8. The drawing shows how the wave of activity spreads from the SAN over the surface of the heart. The figures are in seconds. They show the time for the wave of electrical activity to reach each of the points shown.

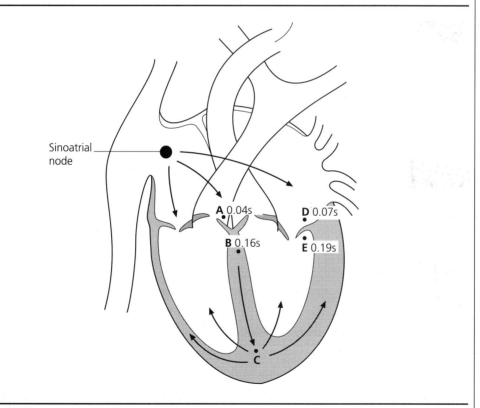

Figure 6.5 *The path taken by the wave electrical activity as it spreads from the sinoatrial node (SAN) over the surface of the heart*

Exercise 6.5 *continued*

1   (a)   The magnification of the drawing in Figure 6.5 is $\times 0.8$. Calculate the actual distance from point **A** to point **B**. Give your answer in mm.

(1 mark)

(b)   Use your answer to question 1 (a) and the times shown on the drawing to calculate the rate at which the wave of electrical activity passes from point **A** to point **B**. Give your answer in mm s^{-1} and show your working.

(2 marks)

(c)   The wave of electrical activity passes very slowly from point **A** to point **B**. Explain why this slow rate is important in coordinating the heart beat.

(2 marks)

2   Explain how you would calculate the rate at which the wave of electrical activity passes from point **B** to point **C**.

(2 marks)

3   Points **D** and **E** are very close together but the wave of electrical activity arrives at point **E** 0.12 s after it arrives at point **D**. Use information on Figure 6.5 to explain the difference in time.

(2 marks)

We can study the electrical changes which take place in the heart by using electrocardiograms (ECGs). Body tissues conduct electricity so the electrical changes which occur as the wave of electrical activity spreads over the heart can be detected on the body surface. When an ECG is made, electrodes are attached to the chest. The signals from these electrodes are processed and displayed to give a picture of the events which occur as the electrical activity spreads over the heart. When we look at this picture we can see what happens when the electrical activity spreads from one electrode to another. Figure 6.6 shows an ECG with the electrodes arranged to show the spread of electrical activity from the atria to the ventricles.

Exercise 6.5 *continued*

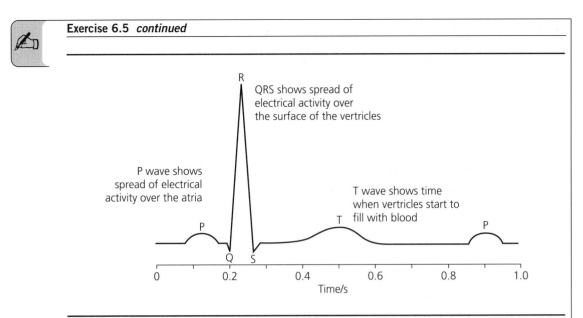

Figure 6.6 *An ECG showing the spread of electrical activity from the atria to the ventricles*

4   Look at Figure 6.6.

(a)   How long does one heart beat take?

(1 mark)

(b)   What is the heart rate of this person in beats per minute? Show your working.

(2 marks)

Hint   This is a very simple question but one that causes difficulties in examinations. If you are struggling, think simple! Put in some very straightforward numbers and work out your method with these. Then use the same method with the numbers you are given in the question.

An ECG can be used to calculate the length of time that the heart is emptying and the length of time that it is filling during a heart cycle. The interval between the beginning of the QRS wave and the peak of the next T wave is when the ventricles are emptying. The interval between the peak of the T wave and the beginning of the next QRS wave is when the heart is filling. Table 6.5 shows the results of an investigation of how emptying time and filling time change with heart rate.

Table 6.5 The effect of heart rate on the emptying time and the filling time of the ventricles

Heart rate/beats minute^{-1}	Emptying time/s	Filling time/s
55	0.43	0.66
60	0.43	0.57
70	0.42	0.44
80	0.42	0.33
90	0.43	0.24

5   (a)   Plot the data in Table 6.5 as a suitable graph.

(4 marks)

(b)   What do the data show about the way in which heart rate increases?

(2 marks)

6.6 Blood vessels and blood flow

In order to answer the questions in this exercise, you must be able to

- describe how blood flows from the heart to the capillaries through arteries and arterioles
- describe how blood returns from the capillaries to the heart through venules and veins
- explain how the blood supply to different organs is affected by exercise.

Here are some remarkable facts. The left side of a human heart pumps out approximately five litres of blood every minute. All this blood passes through the aorta. It would take two years, however, for 1 cm^3 of blood to pass through a single capillary! We can explain this enormous difference in the rate of blood flow by looking carefully at the features of different blood vessels. Some of their features are summarised in Table 6.6.

Table 6.6 Some features of blood vessels and blood flow in a dog

Blood vessel	Total number of vessels	Mean length/cm	Mean diameter/cm	Total cross-sectional area/cm^2	Total blood volume/cm^3	Rate of blood flow/cm^3 s^{-1}
Aorta	1	40	1.0	0.8	30	28
Other large arteries	40	20	0.3	3	60	7.8
Arterioles	4×10^7	0.2	0.002	125	25	1.18
Capillaries	1.2×10^9	0.1	0.0008	600	60	0.036
Venules	8×10^7	0.2	0.003	570	110	0.04
Large veins other than vena cava	40	20	0.6	11	220	1.9
Vena cava	1	40	1.3	1.2	50	1.8

Using the figures in Table 6.6, answer the following questions.

1 Explain why the number of capillaries is greater than the number of arterioles and the number of venules.

(1 mark)

2 Explain how the figures in each of the following columns were calculated:

(a)   the total cross-sectional area.

(1 mark)

(b)   the total volume of blood.

(1 mark)

Exercise 6.6 *continued*

3 Calculate the length of time it would take for a red blood cell to pass from one end of a capillary to the other. Show your working.

(2 marks)

4   (a)   Describe the relationship between the mean diameters of the blood vessels which take blood to an organ and the rate of blood flow through them.

(1 mark)

(b)   There is friction between the blood and the wall of a blood vessel. This frictional force slows down the flow of blood. Use this information to explain why blood flows more slowly in a small artery than in a larger one.

(2 marks)

(c)   Explain why it is an advantage to an organism for blood to flow slowly through capillaries.

(2 marks)

Although all the organs in the body require a blood supply, the rate of blood flow to a particular organ may vary. Figure 6.7 shows the minimum and maximum rates of blood flow through various human organs.

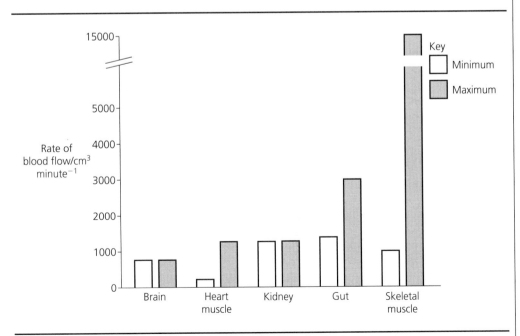

Figure 6.7 *The minimum and maximum rate of blood flow through some human organs*

Exercise 6.6 *continued*

5 Under what body conditions would you expect to find the maximum rate of blood flow in

    (a)   muscle?

    (b)   the gut?

(2 marks)

6 Explain why the change in blood flow to the heart muscle which takes place during strenuous exercise is important.

(2 marks)

7 The total volume of oxygen going to all the organs of the body other than the lungs in 1 minute is 5000 cm^3. Explain why the rate of blood flow to the lungs is also 5000 cm^3 minute^{-1}.

(1 mark)

6.7 Capillaries and tissue fluid

In order to answer the questions in this exercise, you must be able to

- explain the relationship between osmosis and water potential
- describe how tissue fluid is formed and returned to the circulatory system.

A mesentery is a thin layer of tissue between loops of intestine. If you look at a piece of mesentery with a microscope you can see arterioles, capillaries and venules. Figure 6.8 shows a single capillary from a piece of mesentery. We call the pressure that a fluid exerts the hydrostatic pressure. The numbers on this drawing are the hydrostatic pressures of the blood at specific points along the capillary.

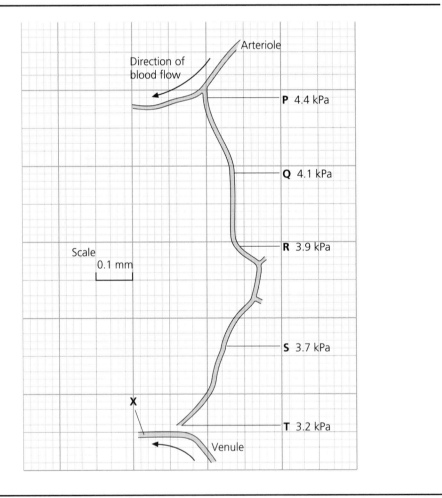

Figure 6.8 *A single capillary from a piece of mesentery*

1 What causes the hydrostatic pressure of the blood in the capillary?

(1 mark)

Exercise 6.7 *continued*

2 Construct a table to show how the hydrostatic pressure differs with the actual distance along the capillary. The drawing has been made on a 2 mm grid to help you make measurements.

(3 marks)

Hint Questions do not often ask you to construct tables so it may be worth looking again at Box 3 on page 15 before you answer question 2.

3 Which of **A**, **B** or **C** below is likely to be the hydrostatic pressure of the blood in the venule at point **X**? Give the reason for your answer.

    A   3.5 kPa

    B   3.2 kPa

    C   2.5 kPa

(2 marks)

Cells are surrounded by fluid which we call tissue fluid. The hydrostatic pressure of tissue fluid is 0.3 kPa. We can calculate the net hydrostatic pressure at point **P** by subtracting the hydrostatic pressure of the tissue fluid from the hydrostatic pressure of the blood. The net hydrostatic pressure at point **P** results in water and the small molecules which are dissolved in it being forced out of the blood into the tissue fluid.

4   (a)   Look at point **T**. On the basis of net hydrostatic pressure alone, would you expect fluid to be forced out of the blood?

(1 mark)

    (b)   Albumin is a soluble protein found in the blood. It has a relative molecular mass of 69 000. Suggest why proteins such a albumin remain in the blood and are not forced out by net hydrostatic pressure.

(2 marks)

    (c)   When you are bitten by an insect, your tissues release a substance called histamine. Histamine causes the walls of the capillary to become more permeable. Suggest why insect bites often result in swelling.

(1 mark)

Exercise 6.7 *continued*

We will now look at another factor which affects the formation of tissue fluid. There is a greater concentration of dissolved substances in the blood plasma than there is in the tissue fluid.

5   (a)   What does this tell you about the water potential of the blood plasma compared with the water potential of the tissue fluid?

(1 mark)

   (b)   How would you expect this difference in water potential to affect the movement of water between the blood plasma and the tissue fluid?

(1 mark)

At point **P**, the tendency of water to move out due to the hydrostatic pressure is greater than its tendency to move back in by osmosis. At point **T**, the tendency for water to move out is less than its tendency to move back in. As a result, water and dissolved substances move out through the capillary walls at **P** and move back in at **T**.

6   Explain why this movement of water and dissolved substances is important to the cells in the tissue supplied by this capillary.

(2 marks)

7   Use information in this exercise and your own biological knowledge to suggest **one** reason for each of the following. As you go further along a capillary

   (a)   the hydrostatic pressure of the blood falls.

(1 mark)

   (b)   the water potential of the blood plasma becomes more negative.

(1 mark)

8   (a)   People who are starving often have swollen ankles due to the accumulation of tissue fluid. Use information given in this exercise to suggest an explanation for this accumulation of tissue fluid.

(2 marks)

   (b)   Some people suffer from high blood pressure. How would you expect high blood pressure to affect the circulation of tissue fluid? Give an explanation for your answer.

(2 marks)